Role of Fisheries in Rural Development

Edited by
S. Giriappa

2022
Daya Publishing House®
A Division of
Astral International (P) Ltd
New Delhi 110 002

First Impression, 1994
Reprinted, 2022
ISBN 9789359191782

Published by : **Daya Publishing House®**
A Division of
Astral International Pvt. Ltd.
– ISO 9001:2015 Certified Company –
4736/23, Ansari Road, Darya Ganj
New Delhi-110 002
Ph. 011-43549197, 23278134
E-mail: info@astralint.com
Website: www.astralint.com

Preface

Rural development has remained an enigma in third world countries beset with population pressure and ineffective resource utilisation. Inadequate increase in productivity and lack of competitive diversification of rural subsectors have posed major threats to rural transformation and advancement aimed at sustainable and symbiotic development process. At the time of the so-called green revolution, efforts were initiated toward a blue revolution to increase fisheries stock and improve rural management. That revolution still is a dream. Sooner or later, it has to be realized so as to initiate the required rural development transformation.

This book is an attempt to evaluate the impact of fisheries sector on the development process of the country. The contribution of marine and inland fisheries in regions like Kerala, Karnataka, Pondicherry, Uttar Pradesh and West Bengal has been evaluated by the eight papers included in the volume.

The editor is thankful to the contributors for their kind cooperation and also to those who had promised but could not contribute due to one or other reason. The assistance of M/s. Gopalakrishna I, Muktha Shet, Rajeshwari and Seetharam M, in editing the papers is gratefully acknowledged. Shri Anil Mittal, Publisher is to be thanked for suggesting the idea of this edition and bringing out the publication in a short time.

S. Giriappa

Contents

Contributors

Ramachandra Bhatta is Assistant Professor of Fishery Economics, University of Agricultural Sciences, College of Fisheries, Mangalore-575002.

Malobica Das belongs to Fisheries Research and Training Centre, G.B. Pant University of Agriculture and Technology, Pantnagar-263145.

Stanley D'Silva is an M. Phil Scholar, Department of Economics, Pondicherry University, Mahe Centre.

S. Giriappa is Professor, SBI Chair in Rural Development, Mangalore University.

P. Ibrahim is Reader, Department of Economics, Pondicherry University, Mahe Centre.

Debabrata Lahiri is Associate Professor, Rural Development Centre, Indian Institute of Technology, Kharagpur.

M. Muktha Shet is a Research Staff in the SBI Chair in Rural Development, Mangalore University.

Mahesh Padhy belongs to Institute of Agriculture, Visva Bharati, Sriniketan, West Bengal.

Kazi M.B. Rahim is Reader, Institute of Agriculture, Visva Bharati, Sriniketan, West Bengal.

C.S. Singh is Professor and Head of Fisheries, Fisheries Research and Training Centre, G.B. Pant University of Agriculture and Technology, Pantnagar-263145.

Katar Singh is RBI Chair Professor, Institute of Rural Management, Anand.

Contributors

Tables

Diagrams/Figures

I
Fisheries Development: An Introduction

S. Giriappa

Conservation demands effective management of the common property resources which are scarce and exhausting. Management of land base, forest, fisheries, water resources, etc, where open access is there inefficient utilisation has resulted in faster depletion of these resources, less sustainment and as a result, environmental degradation. It is evident that even if a ten per cent higher efficient utilization is effected by and through avoidance of wastages, a substantial proportion of these resources can be sustainably conserved. Too much crowding for resource exploitation and lack of competitive prices have resulted in an uneconomical infrastructure, superfluous competition, misidentification of priorities, lack of appropriate technologization, underplay of local interest and several other problems. Proper management of the commons imply more opportunities for the poor and better distribution of power. This is also essential to really decentralize the overall decision making process and avoid over-concentration of various interests.

In rural development which aims at developing the rural areas at par with urban regions, common property management assumes greater significance owing to its specific endowment of the natural resources. For example, in exploitation of water resources, non-competitive prices make water so cheap that this scarce resource is wasted for irrigation purposes. Where irrigation is privatized or competitively priced as in urban areas, the wastages are bound to decline. If community forestry is effectively managed by the community as a whole, deforestation will be minimum than in cases where common forestry is considered as belonging to somebodyelse. In fisheries, overcrowding and the nature of the commodity may force wastage deliberately. If entry is regulated and preservation techniques are sophisticated with competitive prices, it has been pointed out that only half of the present resource exploitation would be necessary to fulfil the existing basic needs.[1]

Total fish production in the world which was around 20 million tonnes in the forties increased to over 110 million tonnes in 1990. In this, the increase in marine fish catch is higher than that of inland source. After the seventies, there has been a diminishing increase in marine fish catch. In the seventies and early eighties, the rate of increase was less than one per cent a year. The main reason for decline in the annual rate of increase has been the limited size of stocks of fish species. Some of these species are already overexploited while others are nearing the limits of capacity. The Food and Agricultural Organisation of the United Nations[2] has estimated that the potential catch of all conventional marine species would be less than 100 million tonnes and by 1990, over 82 per cent of this has already been achieved. This shows the vulnerability of the production system and the necessity to diversify the existing production process beyond the conventional methods. Further, according to the same Organisation[3] most of the potential expansion in marine fish catch would come from increased fishing requiring a huge capital investment in fishing boats and gear which might entail in declining returns per unit of effort for all fishermen. The developed countries account of about 60 per cent of fish catch with a per capita fish consumption of over 30 kgs per annum.[4] Whereas, in some developing countries the per capita consumption could be less than 5 kgs. With the establishment of exclusive economic zones within 200 nautical miles of the shorelines, fish catch is more localised than in the past demanding huge local effort in increasing production of the already vulnerable stock.

Aquaculture, the farming and husbandry of fresh water and marine organisms has been recognised as a potential source of food and industrial product. For the developing countries, it can provide protein as well as foreign exchange. It can also provide alternative employment to fishermen and farmers. Aquaculture production has been increasing rapidly since the seventies. It accounts for about one tenth of combined fresh water and marine capture. Aquaculture is practised to a larger extent in the developing countries, mainly in China. Unlike capture fisheries in which the size of the catch is limited by the size and availability of fish population, aquaculture is limited by the extent of suitable areas. It must be practised close to the shore within estuaries and semi-enclosed bays or in open ocean waters near the shore. The land resource for aquaculture is one dimensional and therefore severe competition amongst the users may ensue. For development of aquaculture and pisciculture, transfer of technology to the developing countries is still in the initial stages.

Unlike agricultural production which can be greatly expanded by the use of suitable technology, productivity of marine organism is limited by the chemical and physical properties of ocean water. The realisation of limited fish stock has necessitated ocean enclosure movement called exclusive economic zone of the coastal countries. The open access characteristics in addition to the limited quantity of resources that can be effectively exploited give rise for a better management. The objectives of such management are to maximise the social and economic benefits of fisheries to concerned marine population and their habitats and to improve international relations thereby. Effective management depends on the availability of relevant data which seem to be lacking in many countries.

Fisheries management has to be improved by developing and adhering to the eco-system conservation principle called unit management wherein all associated species, relevant habitats and spread-outs and necessitated areas are jointly managed. Regulation of catch, entry of vessels, developing endogenous technology for management and provision of marketing and other infrastructures to both capture and artisanal fishermen would improve sustainability and pollution control of the coastal areas. Losses in catch, storage and transport have to be minimised. Improving aquaculture technology would sustain fisheries so as to supplement agriculture and bio-diversity.

II

The collection of papers cover aspects like technology, management and distribution of gains in both marine and inland fisheries. Since fisheries development is crucial for rural and thereby economic development, the papers highlight the economics of different modes of production, conditions of fishing population and resource management.

The specific cases of fisherwomen and cooperatives have also been analysed.

The paper by Ibrahim and D'Silva is on the economics of mechanised boats in the (Pondicherry) Mahe region. They use efficiency measures like catch per trip and effort, labour productivity, capital intensity and cost effectiveness in different crafts-gear combinations like odam-boat seine, thoni, gillnet and trawlers. The thoni-gillnet combination yields a higher net profit than others. Their conclusion is that mechanised boats (trawlers) neither effect a bigger output nor make a larger profit than the motorised crafts. This certainly calls for a shift in the mechanisation policy of the government.

The editor with a brief introduction about the necessity to improve fish production as a supplementary to food production traces the growing importance of fisheries in Indian economy. The growth rate of the sector has been declining in spite of recent developments. This is especially so in the case of states like Karnataka. Taking a case study of mechanised and non-mechanised boats in Dakshina Kannada district of Karnataka, an attempt is made to show the declining share of non-mechanised boats both in quantity and value terms (the latter especially after 1990-91). Considering the operational performance of the boats for different years, it is found that per boat trips have come down, whereas per boat catch has slightly improved. This conclusion would strengthen the viewpoint of selective mechanisation so as to yield a sustainable fish catch.

Ramachandra Bhatta analyses the socio-economic fisheries development in Karnataka. He highlights the inadequate understanding of relationship between developmental programmes and diversity of fishing communities on one hand and inequitable distribution of economic opportunities between mechanised and non-mechanised fishermen on the other. Market fishermen have a greater income potentiality than those under underdeveloped marketing systems. However, the inequitable income distribution has a wider gap within the developed marketing agencies. Overexploitation of marine fisheries resources is the bane of fisheries sector in Karnataka.

Muktha's paper deals with the decision making analysis of fisherwomen in Dakshina Kannada district of Karnataka. After evaluating boat economics, the author assesses work and income contribution by female respondents and their family members in different fishing activities in both mechanised and non-mechanised boat households. The analysis centres around marketing and processing activities of fisherwomen of different age groups in both male-headed and female-headed households endowed with mechanised boats and non-mechanised boats. Work and income contribution by female respondents is found to be significant. In decisions pertaining to work and mobility, females have a significant role but not in households maintenance, account keeping, purchase of assets, education, health etc. However, females in non-mechanised households have a better decision capacity. Her attempt on the impact of technology in different households gives varied results. In the mechanised households, a significant proportion of respondents are aware of the state of technological development and they do adopt the improved methods of production, marketing, transport and storage. However, the proportion of respondents having improved their living conditions because of modern technology is 50 per cent and over 17 per cent have reported deterioration in their lifestyle. In the non-mechanised households, about 40 per cent of them have reported better-off conditions and about 20 per cent deterioration. The analysis shows that mechanisation of fisheries has had mixed implications and it will take a long time to realise its full potentiality.

According to Katar Singh, marine fishermen cooperative societies in the overexploited Kerala coast may yield a promising mechanism in proper resource management and in improving sustainability both in productivity and socio-economic well-being of the fishermen. There has been a significant growth in the cooperative movement at different levels and the cooperatives have been quick in adopting in the fisheries technology. Unionisation among Kerala fishermen has had a beneficial impact on their rights and in cooperative management. The cooperatives have the problems of marketing and lack of support by apex institutions. Exploitation by moneylenders has resulted in limited participation by fishermen in major policy options. Not only the cooperatives have not comeup to expectations also many poor fishermen have been marginalised in the system. In motorised fishing, increased catch has not yielded any increase in net income owing to increased input cost and overdependence by fishermen on external agencies.

Lahiri's paper highlights the important role of fishermen's cooperative societies in West Bengal. Though only about 28% of the PFCs are in working conditions, the cooperative societies are without adequate support and are beset with many problems. The problems vary from exploitation by intermediaries, inadequate credit and input facilities, lack of support by central societies and lack of interaction with local markets. The author calls for leasehold rights of specific water bodies to the PFCs, provision of crop loan, active involvement of the fisheries cooperative federation, improvement in the marketing network for better functioning of the fishermen's cooperatives.

The paper by Rahim and Padhy on the scope and constraints of inland fisheries of Birbhum district of West Bengal analyses subsistence and commercial pisciculture and adoption of improved technologies. The constraints ranging on environmental, technological, socio-economic and infrastructural indicate that fully commercialised fishermen have adopted modern technology and that recommended packages were not uniform across different categories of fishermen. Efficient management, training and marketing are crucial in increased productivity of inland fisheries.

C.S. Singh and M.Das present the prospectives of aquaculture in effecting efficient farm management, operation of fish ponds, site selection, etc. They highlight about the necessity of judicious stock and species selection, fertilizing the pond with manures and other feedstock and the feeding schedule. Following the Chinese example, fish ponds can be a highly profitable venture complementing with agriculture and horticulture, where integration aims at proper utilisation of biowastes and whole farm management.

Fisheries is one sector where lot of research has to be carried out as to its contribution and prospects. The pertinent issues to be tackled are amiable resource management, diversification of production and processing activities and establishing a viable structural mechanism toward development of the poor fishermen and the fishing villages. In this, that the papers should throw some light on the above issues and generate further discussion on the subject.

NOTES AND REFERENCES

1. Repetto, R. *ed. The Global Possible, Resources, Development and the New Century,* Affiliated East-West Press, New Delhi, 1990. Appendix p. 507.

2. FAO, *World Fisheries and the Law of the Sea,* quoted in R. Revelle 'Present and Future State of Living Marine and Freshwater Resources' in R. Repetto *ed op.cit* Chap. 14.

3. FAO, Committee on Fisheries, cited in R.Revelle *op.cit,* p. 443

4. Handbook of Fisheries Statistics 1991, Ministry of Agriculture, GOI, New Delhi, 1991.

NOTES AND REFERENCES

II
Economics of Mechanised Boats and Motorised Crafts

P. Ibrahim
and
Stanley D'Silva

Motorisation of country crafts as a popular strategy to enhance the operational efficiency and labour productivity of traditional fisheries was introduced in the major maritime regions of the country from the turn of the eighties. Motorised craft is viewed as an intermediate technology which is capable of overcoming the inefficiency of the traditional technology on the one hand and the problems created in the wake of mechanised (trawler) fishing-resource depletion, ecological imbalances, deprivation of traditional fishermen on the other.[1] The motorised technology has earned such a wide popularity with the fishermen that in several fishing villages there has been a near total switch over to this technique. The traditional techniques (non-motorised country crafts) are on a steady decline, and in some areas virtually non-existent. The two major techniques that dominate in many major fishing centres now are mechanised boats and motorised crafts. The objective of this paper is to workout the comparative economics of these two techniques.

The study relates to the Mahe region off the Malabar coast. The Malabar coast which forms part of the West Coast of India is one of the most productive in the whole of Indian ocean due to the presence of an expansive continental shelf and the occurrence of such favourable natural processes in the sea as the upwelling of waters, nutrient rich underwater currents etc.[2] Mahe is situated between Mangalore and

Calicut. The most important varieties of fish in this part are oil sardines, anchoviella, cat fish, perches, ribbon fish, mackerels, prawns, white sardines, sharks etc.[3]

The average annual fishlandings in Mahe at present is of the order of 700 tonnes of which motorised sector accounts for 65% and the mechanised sector 35%. Of the total catch sardines constitute 35%, prawns 14% and mackerels 10%.[4]

Mechanised fishing came into Mahe in mid sixties and as in the rest of the Malabar coast it was confined to trawling due to the presence of rich prawn resources.[5] The traditional crafts which largely consisted of dugout canoes came to be motorised by fitting outboard engines from 1983.[6] The growth of motorisation has been so rapid that by the end of eighties all traditional crafts in the area were motorised.[7] The major craft-gear combinations operated in the study area are: (a) motorised big dugout canoe-boat seine (b) motorised small dugout canoe-gillnet (c) mechanised trawlers. The big dugout canoe has normally a length of 9.76 meters, a breadth of 0.915 meters and a depth of 0.76 meters. There were 64 such crafts in the study area. The big dugout canoes normally operate boat seines. The small dugout canoes are 7.32 meters long, 0.915 meter broad and 0.45 meters deep. They operate gill nets and drift nets. Their number was 67. The big dugout canoe costs Rs. 40,000 and the small one Rs. 25,000. The outboard engines fitted in the dugout canoes are Yamaha, Susuki, Johnson and Evinrode, most of them with 8 horsepower. The length of the mechanised boats (trawlers) in use in this area varies from 8.5 meters to 10.2 meters and the power of engine varies from 15 to 45 horsepower. The most predominant gear used by them are trawl nets made of polythene yarn. There were 22 trawlers in the study area.

While launching the investigation, our plan was to cover all the units belonging to the three combinations. However due to the non-cooperation of some of the units, only a few of the units could be actually studied. Table 2.1 gives the distribution of the fishing units actually studied.

The data collected pertain to a single month, December 1988. It has to be mentioned in this connection that fishing being a seasonal operation and as the availability of fish, species composition, catch and prices vary from season to season to get very reliable information data collected have to be spread over the entire year.

Table 2.1 : Distribution of Craft-gear Combination and Selection of Sample Units

Craft-gear Combination	*Number of Units in Operation in the village*	*Number of Units Reported*
Big dugout canoe (Odam)-boat Seine (A)	25	10(40 Per cent)
Small dugout Canoe (thoni)-gillnet (B)	35	7(20 Per cent)
Mechanised trawler (C)	20	6(30 Per cent)

The data for the study were collected in the following way. Each fishing enterprise in this region has one member among the crew who keeps all the accounts of the fishing expedition such as the costs incurred, quantity and value of catch, the share of the catch distributed to each crew member. We entrusted with the accountant of each unit a schedule in which he was asked to enter the information we required. These informations pertained to fishing efforts expended, costs incurred, species, quantity and price of fish caught, earnings of owners as well as other crew members etc. Since the owners in particular and crew members in general of a unit have a tendency to underreport their catch and almost correctly report the catch of others, we have tried to crosscheck these informations obtained so as to make the results as reliable as possible.

In addition we have also administered a separate questionnaire to the owners of the fishing units with a view to obtaining certain other details such as dimensions of crafts, types of nets, costs of fishing equipments, types of ownership, sources of funds, expenditure pattern etc.

Performance of Fishing Enterprises

In fishing the performance of production units are influenced by several factors such as the skills of fishermen, oceanic and climatic conditions, the accessibility of resources in respect of the distance from the shore, fishing depth, degree of aggregation or dispersion of fish, capital intensity and sophistication of technology.[8]

Efficiency

Conventionally the most popular measure of efficiency of fishing operations is the catch per unit of fishing effort. By fishing efforts are meant the manhours spent in making to and fro trips, in casting the net, and in waiting for the fish to get entangled. Fishing effort is the product of total hours spent at sea (from the beginning to the end of a fishing expedition). Catch or earnings per manhour or pertrip are often considered indices of efficiency of fishing units. Efficiency differences existing among different technologies can be understood from Table 2.2.

Table 2:2 Catch (in Physical and Value Terms) Per Unit of Effort and Trip-techniquewise

Technique (Craft-gear Combination)	*Catch per trip (Kg)*	*Value per trip (Rs.)*	*Catch per manhour of fishing effort (Kg)*	*Value per manhour (Rs.)*	*Number of trips with no catch as % of total trips*
A	296	822	2.6	7.2	23.8
B	34	285	1.63	13.58	27.2
C	39	426	1.06	12.64	38.09

Catch per trip is the highest in the odam-boat seine combination (A) both in physical and value terms. In terms of catch per effort the odam-boat seine combination comes first and the trawlers (C) last. In terms of gross earnings per effort thoni-gill net combination (B) holds the first position, trawlers second and odam-boat seine combination the third position. It may be noted that the performance of mechanised boats in value terms is much better than their performance in physical terms. This is because the catch of mechanised boats largely consists of high priced varieties like prawns unlike the other techniques which mainly catch the low priced species. In order to judge the relative efficiencies, it is also necessary to look into the outcomes of trips. Table 2.2 shows that the trawlers account for the highest percentage of trips without catch (38%) and odam-boat seine the lowest (24%).

Factor Productivities

Productivity concepts highly relevant in the context of fisheries are labour and capital productivities. Labour productivity in fisheries is generally expressed as the yearly catch or gross earnings per fisherman, and capital productivity as the yearly catch or gross earnings per capital invested.

Capital cost involved in different enterprises was worked out on the basis of the present resale value of the fishing equipments, as the majority of the fishing equipments in the area were found to be very old. In the case of odam-boat seine combination though a brand new unit costs Rs. 1.5 lakhs the average resale value amounted only to roughly Rs. 85,000. The capital cost for thoni gillnet combination workedout to Rs. 36,000 while a new one would cost Rs. 60,000. The capital investment per trawler was similarly estimated at Rs. 120,000 (a brand new one costs about Rs. 1.5 lakhs). Thus investment per production unit is the highest in trawler. The capital intensity can be more correctly expressed in terms of capital effort ratio and investment per crew member. In terms of either of the indices, capital intensity is the lowest in the odam boat seine combination and highest in trawlers. For the purpose of computing productivity indices we calculated the yearly gross earnings of different techniques by multiplying the monthly earnings of each combination by corresponding period of operation in a year. Capital and labour productivities are shown in Table 2.3.

Table 2.3: Capital Intensity, Labour Intensity and Capital Productivity of Alternative Combinations

Craft-gear Combination	Capital intensity (Investment per crew member)	Labour Productivity (Gross earnings per crew member per annum) in Rs.	Gross catch per crew member per year in Kg.	Capital Productivity (Gross earnings per unit of investment per annum) in Rs.
A	5306	12960	4668	2.44
B	12000	25099	3024	2.09
C	30000	20142	1867.5	0.67

It is obvious from the table that capital intensity was the highest and capital productivity the lowest in the case of trawler. Comparison of columns 1 and 2 of Table 2.3 shows that trawler is the only technique which failed to achieve average yearly gross earnings equal to its capital cost. In terms of labour productivity thoni-gillnet combination comes first (Rs. 25,099), followed by trawler (Rs. 20,142) and Odam boat seine (Rs. 12,960) in that order.

Cost Effectiveness

In most industries, cost is exclusively born by the entrepreneur. Fisheries, on the other hand is characterised by complete or part sharing of the costs by the crew.

Costs, by convention, can be divided into fixed and variable. Under fixed costs, we have included only insurance and depreciation. Depreciation is calculated on straight line basis considering the reported initial investment and the reported average life time of the respective crafts and gear. The ratio of depreciation to gross earnings can be taken as a measure of the extent of output set apart for capital investment. This ratio varied from 4.4% for the odam-boat seine to 8.2% for the thoni-gillnet and 7.4% for the trawlers.

The variable costs (total operational costs) consist of owners' operating costs and common operating costs. In the study area, normally, owners bear only a part of the operational costs, i.e. the repair and maintenance costs of crafts and gears. However, in such exceptional situations as when the catch is very low and the total operating costs exceed gross earnings, the owner himself would bear the entire expenses of a trip. Common operating costs are those which are deducted from gross earnings before apportioning it between the crew and owners. Usually, expenses incurred on food and drinks during and immediately before and after the trip, fuel during the trip, conventional taxes and offerings to religious institutions are treated as common operating costs. In the case of trawlers, batta and food are also normally reckoned as common operating costs.

Table 2.4 presents the cost composition of different craft-gear combinations. Of the three technologies, fixed cost is the highest for the trawlers. This is accounted for mainly by the insurance premium (in Mahe, only trawlers are insured) and the high interest on loans taken by them for the acquisition of production equipments.

Coming to the variable costs, we found that owners' operating costs account for 8% of the total earnings of trawlers, while it is only 2% in the case of the other two combinations. This difference is mainly because of the comparatively larger expenses that the trawlers incur on maintenance and repair of crafts. The common operating costs come to as high as 50% of the total earnings in the trawlers while it is only 25% in the case of other two technologies. In all the three combinations, fuel is the largest single component of the common operating costs. It accounts for 30% of the total earnings of trawlers, 14 to 18% in the other two.

Table 2.4: Cost Composition of the Craft-Gear Combination (Rs.)

Items	*Odam-boatseine (A)*	*Thoni-gillnet (B)*	*Trawlers (C)*
1. Gross earnings	17282 (100)	6275 (100)	8952 (100)
2. Fixed Costs	1004 (5.8)	666 (10.6)	1419 (15.84)
(a) Interest on Loans	228 (1.32)	150 (2.4)	553 (6.17)
(b) Insurance	-	-	200 (2.27)
(c) Depreciation	776 (4.4)	516 (8.2)	666 (7.4)
3. Total Operating Costs	4385 (25.36)	1521 (24.24)	5195 (58)
(A) Owner's operating Costs	347 (2)	125 (2)	725 (8.1)
(a) Repair and maintenance of Crafts	172(0.99)	75 (1.2)	500 (5.58)
(b) Repair and Maintenance of net	150 (0.86)	50 (0.8)	75 (0.85)
(c) Others	25 (0.15)	-	150 (1.67)
B. Common operating costs	4038 (23.36)	1396 (22.24)	4470 (49.9)
(a) Traditional tax and offerings	190 (1.18)	128 (2.03)	120 (1.34)
(b) Materials	-	-	-
(c) Commission	494 (2.85)	192 (3.07)	-
(d) Fuel	3037 (17.58)	876 (13.96)	2727 (30.46)
(e) Ration	-	-	593 (6.6)
(f) Batta	-	-	1030 (11.5)
(g) Others	317 (1.75)	200 (3.18)	-
4. Remuneration to the crew in cash	9167 (53.09)	2177 (34.7)	1612 (18)
5. Remuneration to the crew in kind	550 (3.18)	400 (6.37)	450 (5.06)
6. Net Profit	2174 (12.57)	1511 (24.09)	276 (3.1)

Figures in brackets are percentages.

The commission charges given to 'dallals' for marketing of fish accounts for 3% of the total earnings of odam-boat seine as well as gillnet combinations. Since dallals and merchants of prawns are one and the same persons, this item of cost does not appear in trawlers. An additional item of common operating costs is with respect to batta and food prepared on board. Generally, the crew members of the trawlers are allowed Rs. 10 towards batta and Rs. 5 towards food, per day of operation, these two together accounting for 18% of the total earnings of the trawlers.

It is seen that common operating cost is the highest for the trawlers (Rs. 5195) and accounts for 50% of their total earnings. The corresponding figures for odam-boat seine technology is (Rs. 4305) 25.36% and that for thoni-gillnet combination (Rs. 1521) 24.34%. The use of a single out-board motor and small size of the crew may be the major reason for the comparatively lower operational cost in thoni-gillnet combination.

Remuneration to the crew is viewed as a cost from the owners angle. 53% of the earnings went as crew remuneration in odam-boat seine, 34% in thoni-gillnet and 18% in trawlers.

Profitability

Profit or loss is the residue which remains after meeting all the costs and allocations. Profit as a proportion of the total gross earnings is the highest in thoni-gillnet combination (24.09%), and the lowest in trawlers (3.1%) with odam-boat seine occupying an in between position (12.57%). The quantum of average catch and gross earnings per Rs. 100 of operating costs are also found to be the lowest in the case of trawler (Table 2.5).

Table 2.5 : Average Catch and Gross Earnings for every Rs. 100 of Operating Cost

Craft-Gear Combination	*Catch in Kg.*	*Gross earnings in Rs.*
A	141	394
B	49.7	412
C	14.5	172

Divisible Earnings

Divisible earnings are that part of the value realised from fish sold which are distributed as remuneration in cash to the crew and to the owners i.e. gross earnings from fishing minus the common operating costs. The system of sharing the divisible earnings is based upon certain conventions. Basically there are three shares: one for the crew, one for the craft and one for the gear. The share of the crew can be considered as the returns to labour, the shares for the craft and gear as the returns to capital. It is out of the latter that depreciation and interest charges are often made.

Table 2.6: Sharing Pattern of Divisible Earnings—Combinationwise

Combination	*Crew's Share* %	*Owner's Share* %
A	65	35
B	50	50
C	40	60

Table 2.6 shows that the highest share goes to the owner in the case of trawlers. This is due to the fact that the owner has to make a very large investment in this combination. The owner gets the lowest share in odam-boat seine combination. This is dictated by the large size of the crew in this combination, as otherwise the earnings of the crew members may become extremely low.

The average share that goes to the workers and owners per trip is of course an index of the relative efficiencies of the alternative combinations. However, from the standpoint of both the workers as well as owners what is more important is the total remuneration received from a fishing unit per trip as well as over a period of time (a year). Since in Mahe there exist fishing units individually owned and collectively owned and three groups of people are involved in fishing enterprises namely worker fishermen, owner workers, and absentee owners, income accruing to each of these groups has to be separately worked out. Table 2.7 shows that though divisible earnings per trip was the highest (Rs. 587) for the odam-boat seine combination the earnings accruing to the crew member was the lowest (Rs. 23.8). This may be due to the large size of the crew in that combination. On the other hand the thoni-gillnet combination offered the highest remuneration per crew member (Rs. 32.8) even though the divisible earnings in its case was the lowest (Rs. 197). The percapita remuneration to the worker-fisherman (non-owner worker) was the lowest (Rs. 21) in trawlers as compared to the traditional techniques. However, the percapita remuneration shown in the table does not include two important components namely batta and food which are allowed to the workers in the mechanised boats, catch or no catch. Conventionally the costs incurred on this account are treated as common operating costs. In Mahe, batta allowed is Rs. 10 and food expense allowed is Rs. 5 per day of operation. Thus the remuneration of workers in a mechanised boat would amount to Rs. 36 (21+15) which is higher than that in the other two techniques.

Table 2.7: Earnings Per Trip Combinationwise

Craft Gear Combination / *Items*	A Rs.	B Rs.	C Rs.
1. Total divisible Earnings	587	197	213
2. Earnings of a non owner worker	23.8	32.8	21.3
3. Earnings of an owner worker (individual)	74.8	131.3	149.1
4. Earnings of an absentee owner (individual)	51	98.5	127.8
5. Earnings of owner worker (collective)	36.6	65.6	53.25

The absentee owner got the highest returns per trip (Rs. 127) in the trawler and the lowest in the odam-boat seine combination. Similarly owner worker under individual pattern of ownership got the highest returns per trip in trawlers and the lowest in the boat seine combination. It is significant to note that in all the three techniques the average returns accruing to a non-owner worker in an individually owned unit is considerably lower than what his counterpart gets in a collectively owned and operated unit. This may be one of the reasons why there is at present a trend towards collective ownership in the study area.

In terms of the total remuneration accruing to the members how the different combinations fare? Table 2.8 throws light on this aspect. The table shows that highest remuneration per crew member (Rs. 800) exists in gillnet combination, the next highest in odam boat seine (Rs. 530) and the lowest in trawlers (Rs. 441). But the crew of trawlers get an additional benefit of Rs. 15 by way of batta and food per day of operation.

Table 2.8 : Per capita Remuneration to Workers During the Period of Operation (December 1988)

Craft-Gear Combination	Per capita remuneration in cash		Per capita remuneration in kind (imputed value)		Total	
	Rs.	%	Rs.	%	Rs.	%
A	499	(94.0)	31	(6.0)	530	(100.0)
B	725	(90.6)	75	(9.4)	800	(100.0)
C	402	(91.0)	39	(9.0)	441	(100.0)

Conclusion

Motorised crafts and mechanised boats (trawlers) have come to stay as the major fishing techniques in several fishing villages of the Malabar coast. At the time of the enquiry three techniques (craft-gear combinations) were in operation in the study area; the odam-boat seine combination, thoni-gillnet combination (both belonging to the motorised variety) and the trawler (mechanised boat). The major points emerging from the study of the comparative economics of the mechanised and motorised techniques may now be summed up.

In terms of weight of fish landed per man hour of effort odam boat seine combination comes first, thoni gillnet second and trawler third. The value of catch per manhour was the highest in thoni-gillnet and lowest in odam-boat seine, trawler remaining in between. The net profit rate varied from 13% to 24% in the two motorised traditional techniques while it remained as low as 3.1% in trawlers. Thus contrary to the prevailing notion, mechanised boats neither produce the biggest output nor make the largest profit.

Highest labour productivity was recorded by thoni-gillnet and highest capital productivity by odam-boat seine combinations. In either case trawlers held an intermediate position. This indicates that the operational efficiency of traditional crafts is better as compared to the mechanised boats (trawlers).

Owner's share was found to be positively varying with the capital intensity per worker. It was the highest in trawlers and lowest in odam-boat seine combination. Eventhough profit rate was the lowest, gross earnings and remuneration per crew member were the highest in trawlers. This may explain, why, other things remaining the same, fishermen prefer to work in the mechanised boats.

All the techniques were operating at a profit, though profit rates varied considerably among themselves. Irrespective of the differences in technology, collective pattern of ownership was found to offer better returns to the actual producers of fish. This may partly explain the current tendency among the fishing enterprises of the area to organise themselves under collective ownership.

NOTES AND REFERENCES

1. Achary Thankappan, *The Socio Economic Impact of Motorisation of Country Craft in Purakkad Village, a Case Study,* Fisheries Research Cell, Programme for Community Organisation, Trivandrum 1986.
   Also see, N. Balakrishnan Nair, *Report of the Expert Committee on Marine Fishery Resources Management in Kerala,* Govt. of Kerala, 1989.
2. Ibrahim. P., *Fisheries Development in India,* Classical Publishers, New Delhi, 1992.
   Also see Ummerkutty. A.N.P., "The Sea-Kerala's New frontier of Hope", in Nair P.K.B. (ed) *Development of Kerala Problems and Prospects,* University of Kerala, 1972.
3. Stanley D'Silva, *Fishing Industry of Mahe,* M.Phil dissertation submitted to Pondicherry University (Unpublished), June 1989.
4. *Ibid.*
5. Ibrahim, *Op cit.*
6. *Ibid.*
7. *Ibid.*
8. John Kurien and Rolf Wilman, *Economics of Artisanal and Mechanised Fisheries in Kerala,* FAO/UNDP, Madras- 1982.

III

Performance of Mechanised and Non-mechanised Boats: A Case Study in Dakshina Kannada, Karnataka

S. Giriappa

Introduction

It is evident that to meet the increasing food demand of the teeming millions, the production of non-land and non-crop food items has to be increased. Without which there is bound to be inadequate food supplies and malnutrition globally or regionally. In this, exploitation of fishery resources both inland and marine has great prospects in supplementing the cereal food requirements. There are countries and regions which have a very high percapita fish consumption. Since fish is a perishable commodity, its marketing and distribution are restricted according to the nature of processing, preservation and distribution technology which may influence an even consumption throughout the year.

In a way what the land loses by way of soil erosion etc, becomes a gain to the oceans benefiting the phytoplanktons and zooplanktons to increase in their number and thereby cause an increase in fish population. However, pollution of sea water restricts not only their population but also their nutritive value.

In terms of per capita consumption, fish consumption accounts for over 30% of daily nutritive requirements in countries like Japan, Iceland, Portugal, Norway, Finland, Maldives, etc, and more than 10% of food requirements in Canada, U.S.A., Australia, EEC countries and other coastal countries and states. In India, the average consumption is about 10 gms per day percapita accounting for less than 2% of food requirements, though it is a major fish producing country. This indicates the vast scope for increased dependency on fish consumption. India is able to exploit only about 25% of its potential fisheries sources.

Japan, Russia, China, Peru, USA and Chile are the major fish producing countries in the world. India is the seventh largest producer with 3.6% share of world production. However, in the case of penaeid prawns, India accounts for about 10% of world production. During the eighties, India's share in total fish production has slightly increased from 3.2% in 1983 to 3.64% in 1989; its share in prawn production has declined from 15.7% in 1983 to 9.1% in 1989, indicating the necessity for improved prawn catch.

The fisheries sector in India contributes about 1% of national income and provides employment to about 2 million skilled and unskilled persons in various activities. Within the agriculture sector, the contribution of fisheries is about 3.3%. It may be pointed out that the share of fishery sector was insignificant in the seventies and only after 1981-82 that its importance has been duly emphasised. The country has a vast coastline extending over 800 kms and inland fishery resources covering both fresh and brackish water of about 10 million ha. The increased fish production achieved during the planning era is only a small part of the gap in the potentiality. The percapita fish consumption is only one fourth of the desirable level. If proper mechanisation and deep sea fishing methods are adopted, fish production in India can increase atleast 2-3 times than the existing level of around 3 million tonnes. Fish production has increased in the country by about 3 times in 40 years and has remained more or less stagnant in the eighties. Fluctuation in fish catch is very high which can be seen by the fact the annual growth rate has remained at 5% during 1960-71, declined to 3.9% in 1971-80, further to 2.7% in 1981-86 and only in 1990-1992 that the growth rate seems to be taking off from the stagnancy. Recent activities in fish production like mechanisation, deep sea fishing, aquaculture, prawn farming and improved distribution methods have had limited impact and these have to be further strengthened in order to make the industry viable.

Production Trends

Table 3.1 presents marine and inland fish production during 1984-85 and 1990-91 in different states of India. In marine fish production, share of Karnataka has declined from 9.9.% in 1984-85 to 8% in 1990-91. States like Gujarat and Kerala which are the major fish producers have substantially increased their share during the period. The share of Maharashtra and Tamil Nadu, the other major producers has shown a decline. Marine fish production in the country which was 1.7 million tonnes in 1984-85 increased to 2.3 million tonnes at around 5% growth per annum. In the case of inland fisheries, the increase has been substantial from 1.1 mil tonnes to 1.5 mil tonnes. West Bengal is the major inland fish producer (accounting for over 36%) followed by Bihar, Andhra Pradesh, Uttar Pradesh and Orissa. Karnataka's share is between 3 to 4%. The combined marine and inland fish production during 1990-91 has been 3.84 mil tonnes; West Bengal leads in production (17.7%), followed by Kerala (14.4%), Gujarat (14.2%), Maharashtra (10.1%), Tamil Nadu (9.7%) Andhra Pradesh (6.9%) and Karnataka (6.2%). Karnataka's share in export of marine products is about 4.5%. In value terms, the unit value realised per kg of export has been Rs. 74.3 in Karnataka and Rs. 80.1 in India during 1991-92. These observations indicate that fish production in the state is not progressing as it should be and also its export is not proportional to its production.

Table 3.1: Share of Different States in Marine and Inland Fish Production (%)

State	Marine		Inland		Total
	1	2	1	2	2
Andhra Pradesh	7.8	5.6	9.5	8.9	6.9
Goa	2.1	2.3	0.1	0.2	1.5
Gujarat	16.9	21.7	2.2	2.9	14.2
Karnataka	9.9	8.0	3.6	3.4	6.2
Kerala	21.6	22.4	2.5	2.4	14.4
Maharashtra	19.5	14.1	2.2	2.4	10.1
Orissa	2.7	3.4	4.8	5.4	4.2
Tamil Nadu	15.6	12.6	14.5	5.3	9.7
West Bengal	1.9	1.3	33.5	36.1	17.7
Assam	-	-	4.4	4.9	2.0
Bihar	-	-	10.0	10.4	4.2
Madhya Pradesh	-	-	2.2	2.4	1.0
Uttar Pradesh	-	-	4.5	6.8	2.7
All-India ('ooot)	1689.3	2299.7	1103.2	1536.3	3835.9

1=1984-85; 2=1990-91

Source: Handbook of Fisheries Statistics 1991, Ministry of Agriculture, New Delhi 1991.

The Case Study

An attempt is made in this paper to study the production trends in both the mechanised and non-mechanised fishing sectors in Dakshina Kannada district of Karnataka during eighties. In the district, there are about 85 fishing villages and around 20,000 active fishermen; cooperative movement is marked even in the case of fisherwomen. Mechanised fishing in the district was initiated in the early fifties. The number of mechanised boats in the early eighties was 1600 increasing to over 2500 in 1991-92. Marked increase has been noticed in the case of gillnetters and long-liners. The number of non-mechanised boats has increased from 4400 in the early eighties to 5200 in 1991-92. Among the mechanised boats, trawlers followed by purse-seiners and gillnetters and in non-mechanised, dugout canoes and plank built boats share the bulk. With the growth of mechanisation, an increase in the number of ice plants, freezing plants, cold storage, frozen storages and canning plants has also been registered.

In the initial stages of mechanisation, the contribution of non-mechanised sector in total fish landing was over 85%. During recent years their share has been reduced to less than 7% but in value terms it was significant at over 23% till 1989-90 but during 1990-92, the share has declined to less than 10%. The purse-seiners account for about 75% of fish landing by mechanised boats, the share by trawlers being 18%.

Fish Production in Dakshina Kannada

During 1963-64, total marine fish catch in the district was 0.35 lakh tonnes increasing to 0.66 lakh tonnes in 1970-71. Fish landing was lower in 1975-77 (0.4 to 0.5 lakh tonnes). Fish production in 1979-80 recorded over 1.44 lakh tonnes, the production declining thereafter. During 1985-86, it again increased, declining in the subsequent years. It is only after 1989-90 that a continuously increased fish production is noticed. During 1990-92, the production touched an all time record of about 1.5 lakh tonnes.

The value of fish catch increased from Rs. 384/t in 1970-71 to Rs. 713/t in 1974-75, Rs. 1140/t in 1977-78, Rs. 1946/t in 1982-83, Rs. 2997/t in 1986-87, Rs. 2683/t in 1989-90 and Rs. 4379/t in 1991-92. For the mechanised boats, the value of fish catch was Rs. 2200/t and for the non-mechanised boats Rs. 9400/t till 1989-90 but during 1991-92, it has been Rs. 4150/t for mechanised and Rs. 7550/t for non-mechanised boats. From the viewpoint of landings, mackerel, oil sardine, penaeid prawns account for the bulk of landings. During 1969-70, mackerel accounted for 24% of total fish catch, while the share of oil sardine was 33% and that of prawns 8%. During 1991-92, the share of mackeral has been 9%, oil sardines 10% and prawns 3% indicating a bigger share for other varieties. This pehnomenon is peculiar to this year only. During 1989-90, the share of the three varieties has been 41.2%, 14.3% and 3.6% respectively, the proportion being more or less similar in the eighties. The value of mackerel was Rs. 4441/t in 1985-86 and Rs. 7952/t in 1991-92. The value increase in oil sardine has been from Rs. 959/t to Rs. 3722/t and that of prawns from Rs. 18780/t to Rs. 27527/t in the period.

Major portion of the fish landing is marketed afresh, the proportion being 48% and 24% goes for curing, 4.5% for reduction, 7.3% for freezing, 2.8% for canning, 1.6% for miscellaneous purposes and 11.6% for fish manure.

Fishing contributes about 4% of the district income. Fish industry is one of the major foreign exchange earners and frozen prawn is the chief commodity exported. Bulk of the prawn landings is exported, the value being over Rs. 25 crores per annum.

Operations and Catch

Table 3.2 presents number of operations and fish catch by mechanised boats from 1980-81 to 1991-92. The number of purse-seiners declined from 282 to 241, trawlers remaining more or less constant (1256 and 1258) and gillnetters increasing from 242 to 397 in the terminal year when compared to the base year. Total number of mechanised boats increased from 1802 to 2537 in the period. The per boat trips increased from 168.5 to 180.7 in the case of purse-seiners. The number of trips was lower during 1986-90. In the case of trawlers, per boat trips declined from 101.5 to 86.2. For all mechanised boats, the declining boat trips has been from 110.6 to 82.2. Per boat catch has been 289.6 tonnes per purse seiner in 1980-81, 303.8 t in 1985-86 declining to 186 t in 1988-89 and increasing to 330.6 t in 1989-90 and to 346.3 t in 1991-92. In the case of trawlers, per boat catch increased from 15.2 tonnes in 1980-81 to 19.8 t in 1985-86 and to 38.3 tonnes in 1991-92. But for all mechanised boats but together, per boat catch remained more or less constant. It was 53.4 t in 1980-81 increasing to 62.5 t in 1985-86, but declining to 51.4 t in 1989-90. It increased to 55.3 t in 1991-92.

Table 3.2: Monthwise Catch by PS (Purse-seiners), Tr (Trawlers) and Mechanised Boats in Dakshina Kannada

	1980-81			1985-86			1989-90			1991-92		
	PS	*Tr*	*Mech*	*PS*	*Tr*	*Mech*	*PS*	*Tr*	*Mech*	*PS*	*Tr*	*Mech*
Per boat trips (No.)	168.5	101.5	110.6	163.4	94.8	101.1	141.7	66.7	64.4	180.7	86.2	82.2
Per boat catch (t)	289.6	15.2	53.4	303.8	19.8	62.5	330.4	17.5	51.4	346.3	38.3	55.3
Per trip catch (t)	1.72	0.18	0.54	1.86	0.21	0.62	2.33	0.26	0.80	1.92	0.43	0.67
April	0.86	0.15	0.32	0.60	0.08	0.17	0.22	0.16	0.18	0.48	0.41	0.39
May	1.07	0.06	0.20	0.54	0.15	0.20	0.48	0.16	0.23	0.47	0.37	0.36
August	-	-	-	-	-	-	-	0.23	0.21	0.59	0.34	0.33
September	2.61	0.48	1.48	2.85	0.32	1.40	3.88	0.18	2.66	2.52	0.33	0.97
October	3.61	0.18	1.47	3.38	0.52	1.80	6.10	0.24	4.46	4.64	0.57	1.56
November	2.53	0.12	1.00	2.41	0.20	0.92	3.20	0.50	1.53	2.47	0.48	0.84
December	2.09	0.17	0.79	1.40	0.22	0.54	1.21	0.52	0.67	1.46	0.41	0.54
January	1.60	0.18	0.48	0.98	0.22	0.38	1.03	0.32	0.44	1.16	0.63	0.62
February	1.29	0.16	0.43	1.18	0.21	0.42	0.73	0.28	0.28	0.66	0.60	0.49
March	1.34	0.17	0.47	0.56	0.21	0.10	0.53	0.17	0.21	0.69	0.35	0.36
All Boats/trips	-	36.5		-	31.8		-	24.1		-	21.8	
/catch (t)	-	13.8		-	15.2		-	16.4		-	12.7	
Catch/trip (t)	-	0.42		-	0.48		-	0.68		-	0.58	

Source: Computed from the data provided by Deputy Director of Fisheries, Mangalore.

Table 3.3 : Catch and Value Per Boat in Dakshina Kannada

Catch/boat (t)	*1985-86*	*1986-87*	*1987-88*	*1988-89*	*1989-90*	*1990-91*	*1991-92*
Purse-seiner	303.8	222.2	186.0	209.7	330.8	351.8	346.3
Trawler	19.8	21.1	20.5	20.4	17.5	39.2	38.3
Gillnetter	3.2	2.6	2.3	2.6	0.6	1.4	1.2
Mechanised	62.5	49.3	43.0	48.3	51.4	60.5	55.3
Non-mechanised	1.2	1.1	1.1	1.2	1.3	1.1	1.2
Value of catch/boat (Rs./lakhs)							
Purse-seiners	5.70	5.04	5.05	5.34	5.45	7.36	13.49
Trawler	0.58	0.80	1.00	0.96	0.69	1.37	1.55
Gillnetter	0.14	0.17	0.16	0.14	0.05	0.14	0.13
Mechanised	1.35	1.37	1.52	1.60	1.13	1.67	2.34
Non-mechanised	0.15	0.14	0.14	0.15	0.16	0.08	0.09

Source: Computed from the data supplied by Deputy Director of Fisheries, Dakshina Kannada.

The per trip catch has increased from 1.72 to 1.92 tonnes for purse-seiners, 0.18 to 0.43 tonnes for trawlers, 0.54 to 0.67 tonnes for all mechanised boats. The production accounting season usually starts from April. But the busy season is during the post-monsoon period. Purse-seiners would be out of work during the monsoon months (June to August) though in 1991-92 they have reported catches during these months also. The bulk of the catch by the purse-seiners is between September to December. Trawlers are busy during October to February, March to May being lean season and during June and July they may be out of work. For all the mechanised boats, the per trip catch during April, May and August is more or less constant (around 0.35 tonnes). During September to November, the per trip catch increases to over 1 tonne (for purse-seiners, 4.6 tonnes). In the case of non-mechanised boats, per trip catch is maximum during September, December and January in the case of rampanis and during December for gillnetters and February for longliners.

The average quantity of fish catch per boat trip is 0.1 tonnes during February to August, around 0.2 tonnes during September to January. For all the boats, trips per boat have declined from 36.5 in 1980-81 to 21.8 in 1991-92 and per boat catch from 13.8 tonnes to 12.7 tonnes. However, per trip catch has increased from 0.42 to 0.58 tonnes.

The per boat value of fish catch has been slowly increasing over the years. This is in addition to increase in fish quantity and substantial increase in unit prices. In the case of purse-seiners, the average quantity of fish catch per boat has been 303.8 tonnes in 1985-86, 222.2 tonnes in 1986-87, 186 tonnes in 1987-88 and 209.7 tonnes in 1988-89. The catch increased to 330.8 tonnes in 1989-90, 351.8 tonnes in 1990-91 and 346.3 tonnes in 1991-92. In value terms, per boat catch was Rs. 5.7 lakhs in 1985-86, Rs. 5.04 lakhs in 1986-88, and over Rs. 5.34 lakhs in 1988-90. The value increased substantially to Rs. 7.36 lakhs in 1990-91 and Rs. 13.49 lakhs in 1991-92. In the case of trawlers, the per boat catch has been around 20 tonnes upto 1988-89, 17.5 tonnes in 1989-90, increasing to over 39.2 tonnes in 1990-91 and 1991-92.The value increase has been from Rs. 0.58 lakhs to Rs. 1.55 lakhs per boat during the period. Table 3.3 describes catch and value per boat during 1985-86 to 1991-92.

In the case of gillnetters, the average catch per boat has been around 3 tonnes upto 1988-89 but it showed a drastic decline in 1989-90 (0.6 tonnes). After 1990-91, the catch increased to 1.4 tonnes but not to the initial level. In value terms except during 1989-90, it has been around Rs. 0.14 lakhs per boat.

For all the mechanised boats, the average catch per boat has been 62.5 tonnes in 1985-86 declining to 49.3 tonnes in 1986-87 and 43 tonnes in 1987-88. It increased to 48.3 tonnes in 1988-89, 51.4 tonnes in 1989-90, 60.5 tonnes in 1990-91, showing a decline to 55.3 tonnes in 1991-92. In value terms, the increase has been from Rs. 1.35 lakhs in 1985-86 to Rs. 1.6 lakhs in 1988-89, showing a decline in the intervening years and again increasing to Rs. 1.67 lakhs in 1990-91 and Rs. 2.34 lakhs in 1991-92. In the case of non-mechanised boats, the per boat catch is around 1.15 tonnes with a value of Rs. 0.1 lakhs in 1991-92. It could be seen that the share of non-mechanised boats has declined very much in recent years. Upto 1989-90, the share of non-mechanised boats in value terms was over 20%, whereas the share declined to 10% in 1990-91 and to just 7% in 1991-92.

Conclusion

It is evident that fish production in Dakshina Kannada has shown that though the number of mechanised and non-mechanised boats have increased tremendously, the catch per boat has not shown remarkable increase and fluctuations do result in under-utilisation of infrastructures in some years. Though the share of non-mechanised boats has slided down after the advent of mechanisation, in value terms, the share has been significant until 1989-90. But it has become very low during recent years. Also, a change in the composition of fish varieties caught by the boats shows that the share of varieties other than mackerel, oil sardine and paneid prawns has increased substantially.

Since the mechanised boats account for bulk of the catch, the value per boat catch has increased significantly after 1990-91 whereas for the non-mechanised boats the value has come down owing to catch of low value fish varieties. Deepsea fishing is yet to break grounds and future policy for the mechanised boats seems product diversification (value added products) and long distancing and year-round fish catch. Since non-mechanised boats would not have these advantages, territorialisation and limiting the seasons would have to be adopted. They should also concentrate on brackish and back water fishing. The fluctuating fish production trends across different states (when two different time periods are concerned) show differing shares in catch over years. This aspect of instability (as in the state of Karnataka) is to be minimised through better performance.

REFERENCES

Dept. of Agriculture and Cooperation, Fisheries Division, *Handbook on Fisheries Statistics*, Ministry of Agriculture, New Delhi, 1991.

Giriappa, S. *Rural Development in Action A Case Study of Coastal Areas*, Daya Publishing House, Delhi, 1993, Chap V.

Raut, K.C. 'Fisheries' in M.L. Dantwala *ed. Indian Agricultural Development Since Independence*, Oxford and IBH Publishing Co P.Ltd, New Delhi, 1991, 2nd ed.

Sharma, U. and S.P. Grover. *An Introduction to Indian Fisheries*, B.S. M.P. S, Dehra Dun, 1982.

Srivastava U.K. and M. Dharma Reddy. *Fisheries Development in India—Some Aspects of Policy Management*, Concept Publishing Company, New Delhi, 1983.

IV

Socio-Economic Impact of Fisheries Development in Karnataka: An Analysis

Ramachandra Bhatta

Background

Forty five years after independence the population of India has grown to nearly two and a half times its size at the time of independence. The latest population census counted 843.9 million Indians on March 1991 (Premi, 1991). During the next 35 years it is estimated that an additional 600 million will be added to this number (United Nations, 1991). India's population is projected to exceed 1000 million by the year 2000 when it would constitute 28 per cent of Asia's population or nearly 17 per cent of world population. A large population soon exceeding 1 billion would be affected by our planning strategy and pace of development. During the past 45 years we have managed to raise the supply of food *parri passu* with the growth in population, but this may require much greater efforts in future. Such a growth in population continue to exert increasing pressure on the food production capabilities of the country. Though the food production kept pace with population growth, it has accumulated strain on production, processing and distribution system (Krishnan, 1992). The decline in the productivity and stagnation of total production has been a part of the common scenario in most of the developing countries, where natural resources continue to be important. The recent literature attributes these changes to population growth, market forces, public intervention, technological changes and environmental stress (Runge, 1981, Repetto and Holmes 1984, Ciriacy-Wantrup and Bishop 1975, Bromley and Chapagain

1984, Jodha 1990, Bromley and Cernea 1989). With the process of commercialisation of production system, the infrastructure and physical marketing improvements were taken up by the governments for creating an efficient production and marketing system. The assumption for investing huge sums of money on harbour and market infrastructure was that the creation of efficient production and marketing system is essential for maximising the benefits of mechanisation in the fishing industry. The improvement in the infrastructure was expected to result in lowering of per unit cost of marketing services and increase the competitiveness of the markets. These general programmes to improve the market system initiated by the government could be classified into three categories. The first category of these programmes was aimed at market infrastructure development through the construction of harbours, roads and markets. The second category of programmes related to the economic policy approaches directed towards producers, consumers and market functionaries such as price policies, subsidies etc. The third category was concerned with creating institutional structure for improving marketing system, such as establishment of regulated markets, marketing boards and co-operatives. Among different institutional arrangements co-operatives were identified as the most powerful setup to provide marketing facilities in the field of fisheries.

Marine Fisheries Development During Five Year Plans

Before India's Independence, the contribution of marine fisheries sector was only at subsistence level. Concerted efforts for the development of the fisheries of the country were initiated after the country became independent in 1947 and through the National Plan Schemes (Five Year Plans), the first of which started in 1951. In the first two Five Year Plan Period (1951-56 and 1956-61), the development thrust in the marine fisheries sector was on mechanisation of indigenous craft, introduction of mechanised fishing vessels improvement of fishing implements, establishment of infrastructure facilities for preservation, processing, storage and transportation of fish and fishery products and for landing and berthing of crafts. In the third Five Year Plan (1961-66), which has a plan outlay of 282.7 million rupees amounting to 0.38% of the total plan outlay, and in the subsequent three annual plan years, these programmes were continued with additional mechanised fishing vessels participating in the fishery. In the IV Five Year Plan period (1969-74) progress was made in strengthening of infrastructure facilities by construction of fishing harbours at major and minor ports and further expansion of export trade. During V (1974-79) and VI (1980-85) Five Year Plans the development emphasis was on the diversification of fishing activities, development of diversified fishery products, motorisation of country craft and exploration of diversified fishing products, and exploitation of fishery resources in the deep sea regions. The declaration of 200 miles of Exclusive Economic Zone (EEZ) in the year 1977 was an historic event during this period. Although the overall marine fish production was showing an upward trend through these years, the rate of increase was not significant and the growth rate during the period was only of the order of about 0.83 indicating the emerging overfishing symptoms in the inshore waters and thus suggested the need for the rational management. In the VII Five Year Plan completed in 1990, the development of deep sea fishing received further momentum (Planning Commission, 1992). In the inshore fishery sector, the motorisation of country crafts progressed considerably.

As a result of these development programmes about 4,75,000 active fisherman all along the coast are participating at present. The fishery resources are exploited by about 1,80,096 non-mechanised traditional crafts, 26,171 motorised traditional crafts, 34,571 mechanised boats employing different types of gears varying from simple seiners to modern trawl nets and purse seiners. With an annual production level of 2.2 million tonnes of marine fish and shell fishes during 1990-91 valued at over 8800 million rupees ex-vessel price the sector contributes to 1.5% of the total national income from agriculture, estimated at 5,74,000 million rupees. The fishery sector contributed over Rs. 28,000 million to the national GDP at current prices accounting for 0.95 per cent of the total GDP in 1987. The country exported 133,653 metric tonnes of marine products worth Rs. 5,312 million during 1990, which constituted 15.52 per cent of the total agricultural exports and 3 per cent of the total exports. An interesting study by Dholakia (1992) showed that the direct and indirect effects of an increase in export demand on gross output was the lowest in fishing (Rs. 1.14) but it was the highest when gross value added (Rs. 0.99). The direct and indirect effects are essentially the first round effects which take into account increase in sectors own output and subsequent increases in the input supplying sector's output which in turn set in motion a chain reaction on the output of all sectors. The export potential of the fishery sector measured in terms of import intensity also highlights the importance of this sector. The study by Dholakia used two alternative measures of import intensity: proportion of imported inputs used for export manufacturing and imported units as a proportion of the value of exports. The study indicated that fishing sector used the lowest value of inputs in the export manufacturing (Rs. 190.6 million) and the ratio of imported inputs to exports (0.6 per cent) among the selected ten sectors.

Production Trend

The trend of marine fish production over the years has shown a four fold increase during the last four decades (1951-1990). The marine fish production registered a quantum increase during 1961-70 with a growth rate of 5.9 per cent. In the following decade (1971-80), although the marine fish production showed an overall upward trend, the decade growth rate was only 0.8 per cent indicating the approaching optimal level of exploitation of the resources in the inshore waters (0-50 m). Despite a production of 2.2 million tonnes recorded in 1989 in the decade 1981-90 the growth rate increase of production was not appreciable vis-a-vis the effort put in, confirming that further intensification of exploitation of these resources in a narrow inshore belt would not realise further increase in production.

The increase or decrease in the annual marine fish production of the country, by and large depends on the success or failure of sardine, mackerel and shrimp fisheries. The wide fluctuations in the annual yield of sardine and mackerel are well known and is generally attributed to fishery independent factors, such as spawning success, recruitment strength and environmental factors affecting the resources. However in the case of shrimp fishery, particularly the penaeid prawns which are much sought after by the export trade, the landings have been fluctuating from year to year and the total production is nearing stagnation in recent years. Further the data on CPUE (catch per unit of fishing effort) and other parameters of the coastal shrimp fishery at major port areas such as Karwar, Mangalore, Calicut, Mandapam etc. have indicated that further increase of effort may not yield increased production as the exploitation has reached the MSY (maximum sustainable yield) (Silas, *et al*, 1984).

Utilisation of Marine Fish and Fishery Products

Fish is used for different purposes and in different forms. The major portion of the catch is utilised for human consumption either in fresh or processed form. The inedible fish is converted into fish meal or manure. The exportable varieties are processed and exported. The utilisation pattern is largely influenced by the food habits of the people, level of income, form of disposal and the availability. In India, prior to 1950, 42.7% of the fish production was consumed fresh, 25.9% converted into sundried fish, 24.8% into salt cured and 6.6% was reduced to fish manure. Over the years, the utilisation pattern underwent considerable changes. The consumption of fish in fresh form gradually increased to 71% in 1966 and subsequently decreased to about 44%. At the same time, curing which accounted for about 44% in 1961 reduced to about 30% of the catch. The utilisation of fish in frozen form contributed to only about 2% of the catch in 1961, and at present it accounts for over 6%. The percentage of the catch used for reduction and offal amount to about 15. Thus, at present, about 44% is used as fresh, 31.5% as dry edible, 14.7% as fish meal and 11% for exports.

The pattern of utilisation of marine fish varies from state to state and depends on catch composition (nature of the resource exploited), and availability of facilities for distribution, preservation, processing, storage and marketing. The southern states contribute more to fresh fish than northern states. Maharashtra, Kerala and Tamil Nadu produce about 68% of dry edible fish, while Gujarat, Maharashtra and Karnataka account for 77% of the fish used for reduction. The maritime States of Kerala, Maharashtra, Tamil Nadu, Gujarat and Karnataka contribute to 91% of the total exportable varieties from the country. It is observed that about 50% of the fish is consumed fresh in and around the producing centres, 45% in demand centres located up to a distance of 200 km from the coast and only 5% goes to the centres located beyond 200 km. Metropolitan cities account for 23.2% of total fresh fish consumption, while rural households consume 45.4%. The bulk of the dry edible fish (69.6%) is consumed in rural areas; the consumption in urban and metropolitan cities being only about 27.2% and 2.3% respectively. The distribution of fish from the landing centre is carried out mainly by road and rail transport. About 55% of inter-state flows of fresh fish and 97% of the flows within the state is by road. The average time taken between the catch and purchase by consumers is estimated to be between 20 and 30 hours and for distant markets more than 70 hours by road. As fish is a highly perishable commodity, urgent necessity for developing cold chain system to reach the fish to distant markets and ensuring quality condition, by improving road and rail transport and proper transport containers is to be stressed.

For better utilisation of fish, particularly in the context of export of fish and fishery products, the country established facilities for modern methods of preservation, storage and processing and intensified research on post-harvest technologies. As a result, there are 322 freezing units (ca. 1486 t/day), 66 canning plants (249.7 t/day), 34 fish meal plants (589.5 t/day), 379 cold storages (35943 t/day) and a large number of ice plants, catering to the needs of the fisheries in the country (James and Rao, 1992). A comparison of fish utilisation pattern in the world and in India shows certain variations. While in India consumption of fresh fish is preferred, the world trend is to consume fish in more refined (processed) form rather than in fresh condition. Similarly, a good portion of the catch is converted into fish meal; it had increased from 22.9% in 1961 to over 34% in recent years. In the leading fish producing countries, the trend is more towards industrial use than for human consumption.

The recent price escalation of shrimps and other quality fishes that resulted from the increased export demand coupled with the rising internal demand for all varieties of fish due to the overall increase in money income and the growing consumer preference for fish, encouraged our fishing industry for the speedy expansion of mechanisation and large scale motorisation of country craft. Hence fishing in our country is slowly transforming from a subsistence level to cash crop operation. However the increased tempo of mechanisation, has not correspondingly reflected in the quantum of fish production. The total marine landings still remains just above two million tonnes, though the potential is much higher than that (Srivastava, 1989). Technical advances tend to increase the capital requirements which in turn make the fishing industry (or fishermen) more conscious about their fishing surplus which depends mainly on the trends in input and output markets. Hence it is highly essential to formulate a fish development policy to assure a reasonable price to producers without impeding the interests of the consumers.

Problems of Development of Market Fisheries

The dilemmas of market fishing development have already occurred in most of the countries. Most of the efforts to develop fisheries have focussed on production of food. This emphasis is of course, consistent with efforts to feed hungry people. The problems of development arise fundamentally from constant change in the technical, ecological, and social aspects of the fisheries that stem from the resources, the competition among fishermen. An example of the complexity emerged at a national seminar on fishery development held in India on April 9-11, 1982. The summary paper on marine fisheries noted the divergent opinions about the fishery resources exploitation; lack of data on commercial operations, economics of fishing enterprises, development of foreign markets and port development and the high cost of parastatal businesses (Reddy, 1983). The small-scale fishermen live in rural areas where opportunities for alternative employment is limited. They usually follow family and village traditions that are closely linked to the seasonal cycles of fishing. They usually follow traditions that have existed for generations and they continue to be involved as families in the related activities of catching, processing, and marketing. But the livelihood of these fishermen were disrupted by the introduction of mechanised fishing. The overall difficulties of market fisheries development become clearer by comparison with the primary objectives of agricultural development. They are (1) growth in output and productivity; (2) sustainability of production systems; (3) stability of production and markets; (4) equity among those who own, earn and market; and (5) efficiency in the use of resources, labour and equipment. The second objective of sustainability of the production systems requires conservation of stocks for market fishing. The equity is judged by access to the resources and market place (Royce, 1987). The ease with which production per person from fishing can be multiplied by technical innovations leads quickly to exceeding the productivity of the stock to ruin the livelihoods of traditional fishermen or both. The introduction of technology to increase production per person would be promoted as progress towards the objective of efficiency. But in market fishing, where fishermen lack ownership technical progress leads to overinvestment, to overfishing with damage to the resources, or both. Because of these problems, some fishery developers have advocated promotion only of appropriate technology. Unfortunately, the technology of fishing, processing, and marketing has been changing rapidly, and maintenance of economic efficiency requires continuing change as well as coordinated social and political changes.

Marine Fisheries Development in Karnataka
Some Issues

Marine fishing is one of the major industries in the maritime state of Karnataka, India. The annual quantity of fish landed is around 200,000 tonnes giving an annual value of Rs. 480,554,000 out of which nearly 35 per cent is exported in value terms. The direct contribution of the marine fishing industry to the State GDP is around 2 per cent at current prices. With 4 per cent of the Indian coastline the State contributes around 6 percent of the total marine fish production of the country.

The state's fishery resource potential is estimated at 425,000 tonnes caught from the Exclusive Economic Zone which has an area of 82,000 square kilometre (Baktha, 1983). Similar to the other maritime states of the country Karnataka has effected a large scale mechanisation of fishing craft for increasing the marine fish production. The State initiated a policy for the exploitation of marine resources in the early 1950's by emphasising the importance of artisanal fishermen. However, after the 1960's there was a radical change in the policy approach in favour of mechanisation. In 1966 a scheme for construction and distribution of trawlers was introduced by the Fishery Co-operative Federation on a pilot basis, a decade after Kerala started its attempt to `modernise' the fishery through an Indo-Norwegian Project. The increasing demand for prawn, lucrative international markets and the Government's interest in the promotion of exports gave a further boost to trawling (Kurien, 1991). Later when purse seiners were introduced in 1976 there was an initial increase in total catches but the fish landings mainly shrimps started declining by 1980. In 1986 the CMFRI (Central Marine Fisheries Research Institute) reported that purse seineing had not increased production. Furthermore it was stated that the activities of the 400 to 450 boats along Karnataka coastline of 300 kms were damaging the resource base (Douthwaite, 1992).

The purse seiners seek out shoals of pelagic resources (mackerel and sardine) and encircle them with their nets which have a less than 10 mm square mesh. Before the introduction of this method both the species were caught by the traditional fishermen, through *'rampani'* nets in the operation of which the whole community participated. The mechanised fishing method has also allowed the wealthy to take over a resource that previously benefited everyone in the coastal communities (Datta, 1989). The mechanised fishing boats have to work from a harbour and consequently the income is likely to be concentrated in a few places.

Thus the outcome of this technological advancement seems perverse. Whilst the industry has not been able to exploit its annual potential from its EEZ there are ample signs of over-exploitation in the inshore waters. The average annual landings in Karnataka over the last 10 years is around 200,000 tonnes which is only 50 per cent of the potential in the EEZ area. It is estimated that the annual average Maximum Sustainable Yield (MSY) of shrimps for Karnataka as a whole is around 8000 tonnes while the present catch has already reached this limit. It is also estimated that the number of boats required to harvest this limit is around 500, with each vessel fishing for an average 200 days per annum. However we have around 2000 trawlers operating along the Karnataka coast. Around some of the major fishing ports the actual catch has exceeded the MSY level. In Mangalore the average annual production of shrimp is 1860 tonnes as against the MSY level of 1715 tonnes (Silas, *et al,* 1984).

The increase in population and the consequent increase in domestic and export demand for fish have led to a steady increase in prices. The annual average price index of fish increased to 582.7 in 1987 while the index for all commodities stood at 395.7 with 1971 as base year (Government of India, 1988). As a result despite exploiting the resources beyond the MSY the fishermen still find opportunities for realising positive economic rent and the fishing industry as a whole holds potential for attracting capital and labour.

Resource Use and Development

The development and market infrastructure play a role in shaping the pattern of resource use. It has been shown that the development of competitive market structures would ignore the external costs and therefore establish a rapid rate of resource use (Howe, 1979). On the other hand market structure farthest from perfect competition would produce a smaller output than a competitive market structure. Thus it appears that the monopolist even while ignoring environmental costs and possibly using a high discount rate is more conservative than the competitive markets. Howe (1979) while analysing some of the reasons for the failure of competitive market structures in the optimum natural resource exploitation states that private markets are likely to overlook the value of stocks of the resources and common access to these resources may preclude the establishment of markets for these resources. According to Jodha (1985) the market and institutional development have only accentuated the erosion of the common property resources which are caused by the absence of property rights and similar groups of pressure on the resources. Hence market forces have played only a complementary role in the process. Apart from the above, market forces have tended to be more strongly felt with the advances in communication and transport facilities over the period of time and they have become more conducive to exploitation of the fishery resources. As Platteau (1984) has pointed out the price fluctuations in a traditional local market are likely to be higher than those observed in a national or international market, due to more erratic fluctuations to which supply is subjected in a small restricted area. The rise in fish prices in major port fish markets led to an increase in the average revenue from the fishery and investment in fishing activities became more lucrative. Under such circumstances the fishing effort, and intensity increased closer to the market. The concentration of more fishermen would also result in the reduction of some of the primary costs of fishing such as the cost of acquiring knowledge about the location of fish through multi vessel information net work in a major port fish markets (Wilson, 1982).

Socio-economic Analysis and Fisheries Development

A major rationale for fishery socio-economic analysis is the realisation that development of economic policies in fisheries requires an understanding of suppositional problems faced by the fishermen. Suppositional problems are those related to the social and economic structure of communities in which fishermen work, the nature of resource base, and the likely effects of development projects and programmes (Pomeroy, 1991). Ian Smith observed that by tacitly assuming unlimited resources, governments have permitted large scale industrial fishing on such a scale that contributed to conflicts with traditional fishermen (Smith, 1979). Ian Smith states "too often decision makers have assumed that fishermen are homogeneous and irrational. No generalisation holds across all cases of course but the opposite assumptions disturb reality less" (Smith, 1979). Many studies of specific communities have given description of the socio-economic fishery system obtained through surveys of fishermen and shore-based fishery workers. For example a series of studies concerned with fisheries on the Atlantic coast of North America have been carried out by Pollnac and Poggie (1979) and Apostle, Kasdan and Hanson (1985). These authors were interested in the effects of changing conditions on coastal communities: "it is essential to understand how inhabitants perceive their present day existence.... Do people continue to live in these small villages by choice, or from lack of alternatives?" (Apostle *et al.,* 1985). Panayotou (1985) categorised fishermen and fishing communities according to their degree of geographical mobility and the occupational mobility (likely determined by market factors and development projects). Most of the fisheries economists agree that the socio-economic rationale for fishermen decision making needs to be better understood. For example, Opaluch and Bock Stael (1984) argue for "a change in research focus from the behaviour of fish to the behaviour of fishermen (since).... the fishermen's decision as to effort level is perhaps the most important type of behaviour to be understood" (P. 107). Social scientists have been at the forefront in pointing out the need to expand the horizons of fishery studies to incorporate social and socio-economic considerations. For example, Anderson (1978) in a report to the Canadian government, reviews social aspects of the Atlantic fisheries and argues that in the face of rapidly changing conditions, "the human and non-human sides of the equation require a systematic monitoring...." Perhaps the major compendium of material dealing with the process of acquiring socio-economic information in small scale fisheries has been produced for the U.N. Food and Agriculture Organisation (FAO, 1985). This paper examines the role of socio-economic information in managing fully exploited fisheries (where concerns of employment, efficiency, food supply and demand and distribution of income and wealth are considered paramount) and in developing new fisheries (where improving the standard of living, increasing fish production and expansion of markets for fish are most crucial.

The introduction of development projects through mechanisation and market development programmes into the common property resources such as capture fisheries can cause severe pressure on local, traditional structure. As the number of motorised boats increase the competition with in the different gears and between the locations for the limited supply of fish and crustaceans (shrimp) becomes very intensive. Especially in regions where there are few alternative job opportunities the entry of increasing number of boats exacerbates the intensive pressure on this common but limited resource. Many studies have shown that the returns of traditional boats are significantly low compared to mechanised boats (Panayotou, 1987).

Objectives

1. To study the development of fisheries and its influence on technology adoption. What are the changes that have occurred in the adoption of technology with the development of markets.

2. The study the impact of development and modernisation of fisheries on the socio-demographic profile of the fishermen community.
3. To study the impact of fisheries development on the income sources of enterprises and its influence on the socio-economic life of the fishing communities.

METHODOLOGY

Area of the Study

Dakshina Kannada district of Karnataka has a coastline of 170 kms, with 77 fishing villages. There are 55 notified landing centres and of these 12 are considered as minor ports and 30 major ports. There are about 9000 fishermen households engaged in fishing activity (Government of Karnataka, 1991).

Characteristics of the Fishing Units Studied

The main types of crafts operating along the Karnataka Coast are

(i) trawlers of length varying from 32 feet to 45 feet, and horse-power varying from 35 hp to 115 hp.

(ii) Purse seiners of length varying from 40 to 45 feet and horsepower varying from 100 hp to 120 hp.

(iii) Indigenous crafts which are wooden canoes of either dugout or plank built construction. The length of these boats varies from 15 to 30 feet. Within these indigenous boats some are motorised with outboard engines.

The primary unit of sampling for the survey was fisherman household. This was defined as a family pursuing fishing as its full time (main) occupation. In some cases all the members of the family worked for the same unit (vessel) while in others they did not. In the latter cases information on each unit of the family was collected, but considered as single household. After collecting the data from 300 households randomly throughout the district the data were post-stratified into three categories. The entire region was classified into three categories based on the differences in the levels of development of port fish markets and institutions. The development of cooperative and other infrastructure facilities were also taken into consideration, while dividing the study area into regions.

Owing to the absence of information on the distribution of fishermen households, it was decided to select the primary households in two stages, the first stage being the fishing village, which was within the area of a fishermen cooperative society and the second stage the fishermen households in the area. The selection of the first stage units was done purposively taking into consideration the representative nature of the village, the availability of samples, and the geographic spread. The first stage units constituted the centres from where data on cooperative institutions were collected. A total of 19 such cooperatives were selected from the three taluks initially. In each selected centre the frame for the selection of sample units was prepared locally under the guidance of the local Secretary.

Table 4.1 : Distribution of Samples In the Selected Taluks

	Households		*Cooperatives*	
Taluks	*Population*	*Sample*	*Population*	*Sample*
Mangalore	2366.00	120.00	14.00	5.00
Udupi	4133.00	110.00	20.00	9.00
Kundapur	2639.00	70.00	9.00	5.00
Total	9138.00	300.00	43.00	19.00

The number of households interviewed (Table 4.1) was limited to 300 since the geographic coverage was large and the villages selected were scattered. Analysis of the data was done by post-stratifying the entire study area of three taluks into three distinct locations based on the nature of market infrastructure development and concentration of fishing units. Based on the information available the fish markets of the district were classified into 3 categories; developed, less developed and underdeveloped markets. The salient features of these markets in terms of the infrastructure available is given in Table 4.2.

(i) Developed Market regions: These were the major port fish markets from where the fishing households operate their crafts and sold their catches. There were 3 such markets in Dakshina Kannada District: Mangalore, Malpe and Gongolli. In the text they are classified as type A markets.

(ii) Less Developed Market regions: This category included the minor ports from where fishing households operate their vessels. They were Ullal, Kulai, Mulki, Pollipu, Udyavar, Hangarkatta, Bejadi, Navunda, Tarapathi. These port fish markets were listed as minor landing centres by the Department of Fisheries, Government of Karnataka. These port markets are referred to as type B market categories in the present study.

(iii) Underdeveloped Market regions: These small landing centres lacks basic infrastructure facilities. They included Baikampady, Haleyangadi, Uchila, Maravanthe, Shiroor etc. They were referred as type C markets in the study.

Table 4.2 : Infrastructure Available in Different Market Categories

Facilities	*Market Categories*		
	Type A	*Type B*	*Type C*
1. Credit	Fully Available	Fully Available	Fully Available
2. Marketing	Fully Available	Partially Available	Not Available
3. Input Supply	Fully Available	Partially Available	Partially Available
4. Provision of Cold Storage Facilities	Fully Available	Not Available	Not Available
5. Berthing & Landing Facilities	Fully Available	Partially Available	Partially Available
6. Fuel Supply	Fully Available	Partially Available	Not Available

The major port fish market and the importance of these markets is shown in Table 4.3. It can be seen that more than 90 per cent of the aggregate quantity was landed in these harbours and 90% of the mechanised boats were operating from these ports. The distribution of fishing units in the selected taluks of district is given in Table 4.4. It shows that Udupi had maximum number of fishing units and villages.

Table 4.3 : Importance of Major Port Fish Markets in the District (Figures in percentage for the ports)

		Centres			
	Total	*Mangalore*	*Maple*	*Gangolli*	*Total (%)*
No. of centres	12.00				
Quantity of fish produced	85782.60 (M. tonnes)	30.50	30.50	28.50	89.00
Fisherman Population	59934.00	9.00	15.76	15.83	40.00
Purse seiners	247.00	36.80	21.05	31.57	89.42
Trawlers	1259.00	28.75	38.52	12.23	79.50
Other mechanised	1330.00	12.11	20.90	12.78	45.80
Traditional	5454.00	15.44	34.84	2.80	53.10

Table 4.4 : Number of Fishing Units and Villages

Types of fishing units	Taluks		
	Mangalore	*Udupi*	*Kundapur*
Trawlers	432.00	673.00	154.00
Purse seiners	107.00	62.00	78.00
Gillnets	205.00	38.00	84.00
Other mechanised	214.00	398.00	391.00
Traditional	1853.00	2894.00	707.00
	2811.00	4065.00	1414.00
Total villages (Fishing)	24.00	45.00	17.00

The primary task of any study concerned with fishermen households is the study of the socio-economic conditions which are a prerequisite for the design and implementation of effective policy and programmes for their development (Panayotou, 1985). In order to obtain bench mark information about the socio-economic conditions and standards of living of the fishermen an attempt was made to provide overall picture of the structure, outfits and standards of living of fishing communities. Later, comparison of the socio-demographic profile of the households located in the different regions (of uneven institutional and infrastructure development) was made. The occupational characteristic features classified into fishing, fishing related and other activities was made. Analysis was also carried out to determine the differences in the fishing income and returns, etc. The total number of households surveyed is given in Table 4.1.

Socio-demographic Profile of the Households

The fishing population of the district is unevenly distributed concentrating mainly in the developed port market regions. In 1990 over 40 per cent of the fishing families and 70 per cent of the fishing crafts were reported to have lived in the developed port market centres. The growth of commercial and mechanised fisheries have greatly increased the number of families living in these centres. The majority of these families were however not urban dwellers, but reside in coastal communities throughout the district. Tables 4.5 and 4.6 show the distribution of households talukwise and marketwise as defined above and such characteristics as family size, educational attainment, employment of women, type of the family, etc. It can be seen that though relatively a high percentage of women work in the fisheries sector, this percentage is less in the developed port market categories and increases with the other two market categories (Market B and Market C). Thus, with the development of the market infrastructure, employment opportunities suitable for women and participation of women in fishing and related activities decreases. This is consistent with the general impact of such development. The role of women's employment and income play a central and often primary role in the fishing families. The evidences indicate that prior to the development of mechanisation, households were mainly dependent upon female income in the fishing communities. The expenditure on basic necessities especially food and nutrition was often found to be more closely linked to women's earnings rather than men's earnings where both men and women earn (Agarwal, 1985).

Table 4.5 indicate that the size of the family varied from 7 to 10 depending on the development of taluks with an average size of 8 which is slightly higher than the national average of 7 members per family. Tables 4.5 and 4.6 show educational attainments of households among three taluks and market categories. The general educational attainment of the fishing households was relatively high compared to other communities. The number of households with graduates was higher in Udupi and Mangalore taluks compared to less developed Kundapur taluk. Similarly, the marketwise educational attainment also indicated that market development had apparently opened up opportunities to fishing households to improve their educational qualifications (Table 4.6). The percentage of uneducated households was around 2 per cent of the total sampled households.

Table 4.5 : Socio-demographic Features of the Sampled Households in the Selected Taluks

Features	*Taluks*					
	Mangalore		*Udupi*		*Kundapur*	
	No.	*Per cent*	*No.*	*Per cent*	*No.*	*Per cent*
Total No. of Households	120.00		110.00		70.00	
Average Family size	7.00		7.00		10.00	
HH with Women working	47.00	39.00	66.00	60.00	38.00	54.30
HH with Graduates	42.00	35.00	43.00	39.00	5.00	7.00
HH with Secondary Education	56.00	46.70	45.00	41.00	32.00	45.70
HH with Primary Education	19.00	15.80	19.00	17.30	33.00	47.10
HH Uneducated	3.00	2.50	3.00	2.70	0.00	0.00
HH Joint family	22.00	18.33	30.00	27.30	12.00	17.10
HH Partnerships	43.00	35.80	44.00	40.00	41.00	58.60
HH Nuclear family	55.00	45.80	36.00	32.70	17.00	24.30
Migrants HH	5.00	4.20	6.00	5.50	0.00	0.00
Average Working members as a % of family size	1.00	14.29	2.00	28.57	2.00	20.00

The type of business ownership clearly indicate the feature, unique to the fisheries sector that a large percentage of enterprises are either joint families or partnerships. The percentage of households under partnership tended to be more in the less developed market categories. It is clear that since fishing is a risky business, in the absence of proper state sponsored schemes to reduce risk, fishing households remain as joint families or form partnerships. However, with the development of the market, the accessibility to the state sponsored subsidised insurance schemes, and institutional credit plans to acquire finance, the fishing households located in the developed market tended to be nuclear. The household migration process also shows that a large number of migrated households are found only in developed market categories.

Table 4.6 : Socio-demographic Features of the Sampled Households in the Market Categories

Features	*Market Categories*					
	Type A		*Type B*		*Type C*	
	No.	Per cent	No.	Per cent	No.	Per cent
Total No. of Households	188.00		39.00		73.00	
Average Family size	8.00		8.00		8.00	
HH with Women working	81.00	43.10	20.00	51.30	50.00	68.50
HH with Graduates	68.00	36.20	10.00	25.60	12.00	16.40
HH with Secondary Education	79.00	42.00	17.00	43.60	37.00	50.70
HH with Primary Education	37.00	20.00	12.00	30.80	22.00	30.10
HH Uneducated	4.00	2.00	0.00	0.00	2.00	2.70
HH Joint family	44.00	23.40	7.00	18.00	13.00	17.80
HH Partnerships	70.00	37.00	16.00	41.00	42.00	57.50
HH Nuclear family	74.00	40.00	16.00	41.00	18.00	24.70
Migrants HH	10.00	5.30	0.00	0.00	1.00	1.40
Average Working members as a % of family size	1.42	17.75	2.00	25.00	3.00	37.50

Growth of Fish Business and Occupational Structure

Table 4.7 shows the growth of fish business and allied activities over the years in different market centres. Access to large urban markets increases the business opportunities within and outside the fisheries sector. The Table illustrates that the sampled households pursuing exclusively mechanised or non-mechanised fishing had decreased over the past 10 years. The number of households pursuing non-mechanised fishing had considerably declined from 43 (14.3%) to only 5 (1.67%) during the last decade. On the other hand households carrying out fishing with other diversified occupations like marketing, processing, labour, agriculture, small business and civil services had increased from 28 to 110 during the last decade, showing an increase of 75 per cent. Similarly, 71 households which were dependent only upon

Table 4.7 : Chronological Development of Fishing Enterprises (No. of households)

Occupational Groups	*Before 1982 (before 10 yrs)*	*1982-1987 (between 5-10 yrs)*	*After 1987 (last 5 yrs)*
Mechanised	74	70	54
Mechanised + Other Occupations	28	90	110
Mechanised + Non Mechanised	6	8	4
Mechanised + Non Mechanised + Other Occupations	3	12	17
Non Mechanised	43	8	5
Non Mechanised + Other Occupations	75	97	109
Crewmen & Other Occupations	71	15	0
Total	300	300	300

Note: Other Occupations include fishery related activities like labour, fish marketing & processing and non-fishing activities like agriculture, transportation, petty business, civil service etc.

labour had diversified their activities to other fishing and related occupations and they are no longer labourers. The inshore fisheries typically are seasonal in nature, resulting in periods of serious underemployment among many fishermen. Especially for small-scale fishermen, occupational specialisation may not provide an optimal level of living conditions. The Table indicates that more than 80 per cent of the fishermen obtain at least a part of their income from sources other than fishing. During the 10 years period this pattern of dependence on fishing appears to have decreased over the years.

An examination of the occupational structure in different market categories showed that the majority of the households (68 per cent) located in the developed market (type A) pursue fishing (mech and non-mech) with other occupations (Table 4.8). However, the extent of diversification of business activities with mechanised fishing tend to decline with the lesser developed market categories (Type B and Type C). In underdeveloped market category (Type C) it could be seen that the households tended to combine non-mechanised fishing with other business activities. This variation reflects important differences which exist within the economy of coastal communities and between the market categories. It also indicated that fishing households with higher educational attainment were trying to find employment and income sources from outside the direct fishing activities, in major fishing ports. There are a number of economic advantages to having a diversity of income sources rather than depending on fishing as a sole source of income. Maintaining a diversity in income sources may be crucial to the long term economic survival of the fishing communities. Fishing is an inherently risky operation both physically and economically and where stocks are heavily exploited and also where marketing problems exist due to inadequate infrastructure diversification of economic activities reduces seasonal underemployment.

Growth in the Intensity of Fishing

An examination of the growth of the intensity of fishing as measured by the engine horsepower was made among households operating trawlers and purse seiners. The trawlers were classified into two major groups one in the range of 35-75hp and another in the range of 75-115hp. Similarly, purse seiners were also classified into two, one in the range of 100-110hp and another in the range of 110-120hp. This classification was arbitrary and based on the distribution of sample households.

Table 4.8 : Occupational Structure of Fishing Families In Market Categories

Occupational Groups	*Market Categories*					
	Type A		*Type B*		*Type C*	
	No.	*Per cent*	*No.*	*Per cent*	*No.*	*Per cent*
Mechanised						
Mechanised + Other Occupations	83.00	44.15	10.00	25.64	17.00	23.29
Mechanised + Non Mechanised	4.00	2.10	0.00	0.00	1.00	1.37
Mechanised + Non Mechanised + Other Occupations	11.00	5.85	3.00	7.70	3.00	4.11
Non Mechanised	4.00	2.10	0.00	0.00	1.00	1.37
Non Mechanised + Other Occupations	45.00	23.94	18.00	46.50	46.00	63.00
Total Families	188.00		39.00		73.00	

Note: Other Occupations include fish marketing & processing, labour, agriculture, small businesses, civil service etc

.Table 4.9 : Growth in the Intensity of Mechanised Fishing

Group	*No. of units at present*	*Before 1980*	*Between 1980-85*	*After 1985*
Trawlers				
(a) 35 - 75 HP	59.00	22.00	47.00	59.00
(b) 75 - 115 HP	85.00	26.00	61.00	85.00
Purse seiners				
(a) 100 - 110 HP	52.00	15.00	40.00	52.00
(b) 110 - 120 HP	56.00	15.00	38.00	56.00

Table 4.10 : Fishing Intensity in Market Categories

Group	No. of units at present	Market Categories: Type A	Type B	Type C
Trawlers				
(a) 35-75 HP	59.00	37.00	9.00	13.00
(b) 75-115 HP	85.00	78.00	3.00	4.00
Purse seiners				
(a) 100-110 HP	52.00	42.00	4.00	6.00
(b) 110-120 HP	56.00	46.00	7.00	3.00

Table 4.11 : Distribution of Mechanised Fishing Units in the Selected Taluks

Group	No. of units at present	Taluks: Mangalore	Udupi	Kundapur
Trawlers				
(a) 35-75 HP	59.00	28.00	25.00	6.00
(b) 75-115 HP	85.00	55.00	22.00	8.00
Purse Seiners				
(a) 100-110 HP	52.00	21.00	24.00	7.00
(b) 110-120 HP	56.00	38.00	12.00	6.00

Total no. of Trawlers

(a) 35-75 HP = 59

(b) 75-115 HP = 85
Total = 144

Total no. of Purse Seiners

(a) 100-110 HP = 52

(b) 110-120 HP = 56
Total = 108

Table 4.9 shows a general increase in the intensity of fishing in both the types of gear over the past 10 years, between the market categories and between the taluk centres. The number of trawlers with higher horsepower (75-115 hp) increased from 26 to 85 representing an increase by more than 3 times during the last 10 year period. Similarly,

Table 4.12 : Occupational Sources of Gross Income in Market Categories

Households	*Market categories*								
	Type A			*Type B*			*Type C*		
	Average Income (Rs.)	*No.*	*Per cent*	*Average Income (Rs.)*	*No.*	*Per cent*	*Average Income (Rs.)*	*No.*	*Per cent*
1. Mechanised	1063589	139	73.94	760575	21	53.80	275774	26	35.62
2. Non Mechanised	184891	49	26.06	126950	18	46.20	124614	47	64.38
Total Families	834567	188		468133	39		178452	73	

the purse-seiners of higher horsepower increased by nearly 4 times during the decade 1980-1991. The intensity of different fishing crafts across different market regions is shown in Table 4.10. The table shows that gears of greater fishing capacity tended to concentrate near the developed market categories. It can be seen from the table that majority of the trawl (78 out of 85) and purse seine gears (46 out of 56) with higher horsepower are located in the major port market region. Thus development of major port fish markets by creating accessibility to the expanded markets which are linked with modern transport and communication system encouraged the growth of more powerful boats with higher power. The massive expansion of modern fishing technology due to official encouragement and also due to market development led to the energisation of fishing craft and increase in the fishing intensity over the years. In other words, within a period of 10 years there was almost 3-4 fold increase in the engine horsepower of the fishing boats. Interestingly it was observed that largest proportion of high intensity fishing crafts were found in the developed port fish market categories. It must be mentioned here that the expansion of greater intensity boats was more rapid during the last 5 years. The present structure of market and property rights encourage the mechanisation of fishing crafts. One reason why it is important to recognise this factor is that the intensity of fishing started increasing only after the catch per unit of fishing effort started declining. The profitability of higher intensity of fishing crafts have been greatly enhanced by various government policies. These include subsidies provided to buy the engines and fuel and liberal financial assistance. Obviously this would polarize the access to fish resources between the fishermen of different regions leading to inequality in income distribution. The distribution of mechanised fishing gears between the taluks (Table 4.11) illustrates that a large percentage of boats, with higher engine power, tend to be based in the developed locations (taluks).

Table 4.12 shows that 74% of the families in the developed markets (Type A) are engaged in mechanised fishing compared to only 54% and 35% in the other two market categories. Similarly, the percentage of the non-mechanised fishing households tended to increase in the less developed (Type B) and underdeveloped market centres (Type C). Also the average gross income of both the fishing groups decreased with the decrease in the level of the market infrastructure (Market B and Market C). Thus the number and percentage of mechanised households and their average income was greater in the developed markets. However, in the case of non-mechanised households the average income was more in the developed markets but percentage of households depending on non-mechanised fishing was less.

Table 4.13 shows some of the reasons for which the sampled households expanded their activities. Among other reasons, profitability and family welfare were found to be the primary reasons, followed by the desire 'to achieve growth' and the potential 'opportunities with the development of new technology'. Thus profitability and technological development have been the main criteria for the development of fishing activities.

Table 4.13 : Reasons for Expansion of Fishing Business
(Figures in number of households)

Indicators	*Ranks (priority)*		
	1	*2*	*3*
Profitability	69	55	25
To Achieve Growth	20	74	13
Family Welfare	103	71	38
Development of New Technology	34	13	0
Demand from the Family Members	19	20	9
Catch/Price Uncertainty	24	3	2
Integration of Activities	1	1	1
Increasing the size of Family	17	18	19
Others	13	17	5

Conclusion

There are a number of critical gaps in the knowledge of socio-economic features of fishing communities. These gaps include inadequate understanding of the relationship between developmental programmes and the diversity of fishing communities. The government initiated efforts to promote changes through technological innovations and development of markets had profound impact upon the fishing communities. For example it is obvious that every fishermen could not gain access to improved technologies and established market structures. Such developmental programmes have led to competition for a limited resource over the years in terms of increase in intensity and had resulted in the over-exploitation of fishery resources around the developed port market areas. Within the fisheries sub sector itself competition between mechanised and non-mechanised fishermen has led to considerable difference in the income in favour of mechanised fishing households. The national developmental policies call for equitable distribution of economic opportunities. However the results of the present study indicate that the government programmes have promoted heterogeneity in the socio-economic characteristic of the once homogeneous community. The results showed that the fishing units located in developed market regions were receiving 3-4 times greater income than the fishing families located in the underdeveloped market categories following same technology. Within the market regions also the average income of the mechanised fishing groups were 5 times greater than non-mechanised groups. Whereas, this difference was only two times in the underdeveloped market categories. This growing inequality of income based on infrastructure development and technology led to the sharply divided fishing community in Karnataka; a minority of prosperous fishermen and a majority of non-mechanised fishermen who are alienated from these means of production. Inequality and over-exploitation are thus two ugly faces of development of technology and market in Karnataka. These two aspects of the current crisis are ultimately inseparable in so far as their common cause lies in the anti-social appropriation of fishery resources by a small minority of mechanised fishermen.

REFERENCES

Agarwal, Bina, 1985. Poverty and Agricultural Growth. *Journal of Peasant Studies*, 13: 4.

Anderson, R. 1978. The Need for Human Sciences Research in Atlantic Coast Fisheries. *Journal of Fisheries Research Board of Canada*.35: 1031-49.

Apostle R., L. Kasdan and A. Hanson. 1985. Work Satisfaction and Community Attachment among Fishermen in Southwest Nova Scotia. *Canadian Journal of Fisheries and Aquatic Sciences*.42: 256-67.

Baktha, N.P. 1983. Development of Infrastructure Facilities for the Optimal Exploitation of the Exclusive Economic Zone. In *Fisheries Development in India*. Concept Publishing Company, New Delhi.

Bromley, D.W. and D. Chapagain. 1984. The village against the centre: Resource Depletion in South Asia. *American Journal of Agricultural Economics*. 66 (5): December.

Ciriacy-Wantarup, S.V. and Bishop, R.C. 1975. Common Property—A Concept in Natural Resource Policy. *Natural Resources Journal*. 15: October.

Datta, K.K., S.S. Dan, and A.K. Datta. 1989. Productivity, Profitability and Income Distribution in Capture Fishery: A Case Study of the Orissa Coast. *Economic and Political Weekly*. 52: A-181-188.

Dholakia, R.H., B.H. Dholakia and Ganesh Kumar. 1992. Issues in Strategy for Export Promotion: An Inter-Industry Analysis. *Economic and Political Weekly*. November M-149-M-157.

Douthwaite, R. 1992. *The Growth Illusion*. Green books, Bideford.

FAO. 1985. Report of the Expert Consultation on the Aquisition of Socio-economic Information in Fisheries. FAO Fisheries Report No. 344 Rome.

Government of India. 1988. Handbook of Fishery Statistics. Ministry of Agriculture, Fisheries Division, New Delhi.

Government of India. 1990. National Accounts Statistics. Central Statistical Organisation, New Delhi.

Government of India. 1992. Handbook of Fishery Statistics, 1991-92. Ministry of Agriculture, Fisheries Division, New Delhi.

Government of Karnataka. 1991. Statistical Bulletin on Fisheries, 1989-90. Directorate of Fisheries, Bangalore, Karnataka.

Howe, Charles. W. 1979. Natural Resource Economics. John Wiley and Sons, New York.

James, P.S.B.R. and P.V. Rao. 1992. Food from the Sea: Present Status and Need for a Technology Mission for its Harvest and Utilisation. Central Marine Fisheries Research Institute, (ICAR), Cochin.

Jodha, N.S. 1985. Market Forces and Erosion of Common Property Resources. In *Agricultural Markets in the Semi Arid Tropics,* proceedings of an International Workshop, October 24-28, 1983, ICRISAT Centre: Patancheru (AP) India.

Jodha, N.S. 1990. Rural Common Property Resources: Contributions and Crisis. *Economic and Political Weekly*. 26: June A-65-79.

Krishnan, T.N. 1992. Population, Poverty, and Employment in India. *Economic and Political Weekly*. November 14: 2479-2498.

Kurien, J. 1991. Running the Commons and Responses of the Commoners Coastal Overfishing and Fishermens' Actions in Kerala State, India. Discussion Paper 23. United Nations Research Institute for Social Development, Geneva.

Opaluch, J.J. and N.E. Bock Stael. 1984. Behavioral Modelling and Fisheries Management. *Marine Resource Economics*. 1: 105-15.

Panayotou, T. 1985. Production Technology and Economic Efficiency: A conceptual Framework p. 95-100. In T. Panayotou (ed.) *Small-scale Fisheries in Asia, Socio-economic Analysis and Policy,* International Development and Research Centre, Ottawa.

Panayotou, T. and Songpol Jetanavanich. 1987. The Economics and Management of Thai Fisheries. *ICLARM Studies and Reviews,* 14: 82.

Planning Commission. 1992. Eighth Five Year Plan 1992-97 and Other Five Year Plans. Government of India, New Delhi.

Platteau, J.P. 1984. The Drive Towards Mechanisation of Small-scale Fisheries in Kerala: A Study of the Transformation Process of Traditional Village Societies, *Development and Change*. 15, (1): 65-103.

Pollnac, R., and J. Poggie. 1978. Social Desirability of Work and Management Among Fishermen of Two New England Ports. Anthropological Working Paper No. 5, International Centre for Marine Resource Development, Kingston, Rhode Island.

Pomeroy, R. 1991. Small-scale Fisheries Management and Development: Towards a Community Based Approach. *Marine Policy,* January, 39-48.

Premi, K. Mahendra, 1991. India's Population: Heading Towards a Billion—An Analysis of 1991 Census Provisional Results, B. R. Publishing Corporation, New Delhi.

Reddy, M.D. 1985. Highlights of the discussion on EEZ and new opportunities. In *Fisheries Development in India: Some Aspects of Policy Management,* p. 529-537. Concept Publishing Co., New Delhi.

Repetto, R. and Holmes, T. 1984. The Role of Population in Resource Depletion in Developing Countries. *Population and Development Review,* 9: 4.

Royce, W.F. 1987. *Fishery Development.* Academic Press Inc., London.

Runge, C.F. 1981. Common Property Externalities: Isolation, Assurance and Resource Depletion in Developing Countries. *Population and Development Review.* 9: 4.

Silas E.G., M.J. George and T. Jacab. 1984. A review of the Shrimp Fishery of India—A Scientific Basis for the Management of the Resources, p. 83-113. In J.A. Galland and P.J.P. Rothschild (eds.) *Penaeid Shrimps, their Biology and Management.* Fishing News Books Ltd., London.

Smith, I.R. 1979. A Research Framework for Traditional Fisheries. ICLARM Studies Report, (2): 40.

Srivastava, U.K. and K. Chidambaram. 1989. Marine Fisheries Development: An outlook at turn of the Century and Strategy for Exports. *Journal of Agricultural Issues:* 1: 93-110.

United Nations. 1991. The Sex and Age Distribution of Population—The 1990 Revision, Department of International Economic and Social Affairs, Population Studies No. 122, United Nations, New York.

Wilson, J.A. 1982. The Economical Management of Multispecies Fisheries. *Land Economics.* 58:414-33.

V

Impact of Fisheries Technology on Fisherwomen

M. Muktha Shet

Introduction

Women, especially in rural areas have remained backward for long due to various socio-economic constraints. These constraints emerge from illiteracy, traditional values and norms, dominant position of males, superstition, economic dependence of women on men, social evils like dowry and polygamy, unaccountability of males, low work participation and discrimination. Though women have been denied equal opportunities all over the world for personal growth and social development, in India it is still worse because of the sex-segregated character of society, poverty and traditional value systems.[1]

The most important and relevant concept in examining the status of women is social change. Social change occurs in all societies as a continuous process which is manifested by alterations in social relationships.[2] The status role of rural women would be different from that of urban-women owing to differing social stratification existing in rural areas. Hence social change should be such that the status of rural women is improved. Khan and Aysha point out three important instruments of social change, namely education, social organisation and social legislation.[3]

Labour participation in different activities both in household and work conditions may be influenced by the nature of attitude females show towards family and outside members. Motivation also plays a role in shaping their attitude formation. Depending upon the role of individuals which is structured by the nature of labour participation, status may be outlined if decisions on work, family and reference groups[4] can be evaluated.

Objective

The problem of women development varies according to the given situation. There are certain sectors like fisheries where modernisation is of recent origin and hence, lots of changes are bound to occur in the role and status of fisherwomen. Unlike in agriculture or household industry, women's participation in fisheries is subject to many peculiarities: (i) they are not involved in the production or catching stages, (ii) their involvement is only in processing and marketing, (iii) their problems therefore will be influenced by the nature of participation of male members in these activities. Where a particular activity signifies a major role for women like fish marketing, the influence of men over them may be found. When both men and women have equal role in certain activities like processing, women may be relegated to traditional methods or given in charge of simpler methods of modern processing, Diversification of employment opportunities, education and training and specialised skill formation have been found to improve their conditions towards modernisation.

Because of the peculiarities fisherwomen face, an attempt is made to study their participation and decision-making roles under different levels of technology.

Sample Description

Of the two coastal districts in Karnataka, Dakshina Kannada exhibits a higher level of production and development in fisheries when compared to Uttara Kannada. It is appropriate to consider this district and also Mangalore region which is the most developed in the district. Four villages namely Bengre and Kudroli having a larger number of mechanised boats and Hoigebazar and Ullal having a large number of non-mechanised boats have been selected for the study. The number of mechanised households selected in the four villages is 19 in Bengre, 16 in Kudroli, 7 in Ullal and 6 in Hoigebazar. Of these households, there are 13 households headed by females.

Households owning non-mechanised boats covered number 52; 6 and 9 in Bengre and Kudroli; and 18 and 19 in Hoigebazar and Ullal respectively. The proportion of female-headed households is higher at 79 per cent in the case of non-mechanised households. In Hoigebazar and Ullal, fisherwomen cooperative societies are also studied. The number of households is 10 in mechanised cases and 15 in non-mechanised cases in these villages.

The reference period has been 1990-91 fishing season starting from April 1990 and ending with March 1991.

Most of the respondents are engaged in marketing, 30 in mechanised households and 36 in non-mechanised households. Among the sample households, 85 per cent of women are engaged in fishing. In the rest, 14 respondents in mechanised and one respondent in the non-mechanised group are not engaged in any sort of work. The proportion of respondents engaged in fishing is highest in Ullal (85.7 per cent), followed by Kudroli (75 per cent), Hoigebazar (66.7 per cent) and Bengre (63.2 per cent) in the mechanised group. All the women except one in Ullal are engaged in fishing in the non-mechanised group.

Of the women respondents engaged in fishing, 84.4 per cent are full-timers and 15.6 per cent part-timers. For the mechanised group, the proportion is 76.1 per cent (highest in Hoigebazar followed by Ullal and Kudroli) and in the non-mechanised group 92.2 per cent (highest in Kudroli followed by Hoigebazar and Ullal). In Bengre, the proportion is lowest for both the groups.

The average family size for the sample is 7.2, male members 2.58, females 3.02 and children 1.6. In mechanised households, average family size is 7.77. In non-mechanised households, average family size is 6.67 and in mechanised households, 7.52 in Bengre, 7.81 in Kudroli, 7.57 in Ullal and 8.66 in Hoigebazar. In non-mechanised households, the average family size is 5.83 in Bengre, 7.11 in Kudroli, 6.5 in Ullal and 6.89 in Hoigebazar. The average number of family workers for the entire sample is 3.41, in mechanised households 3 and in non-mechanised households 3.78.

Majority of sample respondents have education upto primary and above primary level and only 9.27 per cent of them are illiterates. About 58 per cent of them have above primary school education and 11.25 per cent of them have college level education. The proportion of those having higher education in mechanised households in 19.59 per cent, and in non-mechanised households it is 14.78 per cent. In mechanised households, female illiteracy proportion is 7.52 per cent and in non-mechanised households, higher at 10.65 per cent.

Boat Income

Table 5.1 describes the income pattern for 6 groups of sample fisherwomen. The I group (purse-seiner) has an annual income of Rs. 4,04,230; the II group (trawlers) has a gross income of Rs. 2,53,938. The net income of these two groups Rs. 30,811 and Rs. 31,513 respectively after deducting boat expenditure. In mechanised households, the fish sale income is Rs. 8,584 in I group and Rs. 9,315 in II group. The average household income for the mechanised households including all activities is Rs. 44, 295. The average income for the purse-seiner household is Rs. 41, 274; in the trawler household, the average income is Rs. 47,864.

In the non-mechanised households, the average income for the groups is Rs. 18,682, Here, the I group (those working under purse-seiners) has a gross income of Rs. 24,804, the II group (those working under trawlers) Rs. 23,681 and the III group (engaged in other activities) Rs. 13,353. Of this, net income is Rs. 16,140 in the I group, Rs. 14,677 in the II group and Rs. 8,118 in the III group respectively. Income from other activities is Rs. 2,832 in group I, Rs. 2,815 in group II and Rs. 11,623 in group III. In the non-mechanised households, the III group is having an higher income in other activities and their fishing income is low when compared to other two groups, because in the III group, males are engaged mainly in other activities and only women are engaged in fishing. In the I and II groups, most of the male and female members are engaged in fishing. The average net income from all activities for the I group is Rs. 18,972, and in the II group it is Rs. 17,492 and in the III group Rs. 19,742.

Boat Expenditure

Table 5.2 describes the expenditure pattern of different boats. The average operational expenditure of a purse-seiner is Rs 3,73,419; of this, diesel and oil account for Rs. 1,14,395, while wages account for Rs. 1,20,980, repairs and equipments Rs. 46,557, insurance payment Rs. 7,472, loan repayment Rs. 55,961, depreciation Rs. 23,384, and miscellaneous expenditure Rs. 4,667. The average operational expenditure for a trawler boat is Rs. 2,22,424. Expenditure on diesel and oil is Rs. 76,634, wages Rs. 73,815, repairs and equipments Rs.

Table 5.1: Income of Mechanised and Non-mechanised Groups from Fish and Other Activities

Village	Boat size (in feet)	Engine size (H.P)	Age (years)	Boat cost (Rs)	Gross fish income (Rs)	Net fish income (Rs)	Fish sale income (Rs)	Other activity (Rs)	Total income (Rs)
Mechanised									
Purse-seiners									
Bengre	43.4	98.3	6.1	5,63,636	3,95,000	30,231	9,477		39,708
Kudroli	47.1	107.0	5.8	6,50,000	4,17,857	28,724	9,642	2,571	40,937
Ullal	47.5	102.8	6.0	5,00,000	3,91,250	27,672	4,725	1,716	34,113
Hoigebazar	48.8	105.3	6.5	6,50,000	4,18,750	39,203	8,137	6,000	53,340
Average	45.8	102.4	6.1	5,90,384	4,04,230	30,811	8,584	1,879	41,274
Trawlers									
Bengre	37.7	94.3	5.4	3,42,857	2,51,129	23,485	4,124	10,800	38,409
Kudroli	37.7	94.2	4.1	3,16,427	2,55,847	41,865	10,260	3,466	55,591
Ullal	32.7	71.0	6.0	3,01,666	2,49,998	25,367	11,433	6,000	42,800
Hoigebazar	34.5	94.0	4.7	5,75,000	2,62,500	26,268	22,650	9,600	58,518
Average	36.5	90.5	4.9	3,51,051	2,53,938	31,513	9,315	7,036	47,864
Non-mechanised									
Bengre	-	-	-	-	26,050	-	17,084	-	17,084
Kudroli				16,666	23,166	-	19,000	-	19,000
Ullal	-	-	-	2,500	18,000	-	11,175	6,726	17,901
Hoigebazar	-	-	-	14,000	35,925	-	21,805	-	21,805
Average	-	-	-	6,484	24,804	-	16,140	2,832	18,972
II Group									
Bengre	-	-	-	11,666	25,600	-	14,734	2,00	16,734
Kudroli	-	-	-	12,500	22,162	-	13,787	4,800	18,587
Ullal	-	-	-	12,000	23,645	-	17,145	3,744	20,889
Hoigebazar	-	-	-	-	23,766	-	13,186	1,125	14,311
Average	-	-	-	8,055	23,681	-	14,677	2,815	17,492

Table 5.2: Expenditure Pattern for Mechanised Boats

(in Rs)

Village	Diesel & Oil Value	Wages	Repair & equipments	Insurance	Loan repaid	Depreciation	Miscell-aneous	Total Expendi-ture
Purse-seiners								
Bengre	1,16,605	1,18,500	50,590	7,618	46,363	22,545	2,545	3,64,768
Kudroli	1,09,711	1,24,285	52,857	7,285	59,428	25,143	10,428	3,89,137
Ullal	1,12,365	1,17,375	40,000	6,750	65,500	20,000	1,587	3,63,578
Hoigebazar	1,18,547	1,25,625	31,000	8,125	66,750	26,000	3,500	3,79,547
Average	1,14,395	1,20,980	46,557	7,472	55,961	23,384	4,667	3,73,419
Trawlers								
Bengre	77,574	73,928	26,000	5,571	30,857	13,714		2,27,644
Kudroli	79,732	71,785	20,000	5,000	21,380	12,657	3,428	2,13,982
Ullal	74,982	75,000	27,000	5,250	28,000	12,066	2,333	2,24,631
Hoigebazar	64,982	78,750	27,500	6,500	31,000	23,000	4,500	2,36,232
Average	76,634	73,815	24,105	5,407	26,929	14,041	2,105	2,22,424

24,105, insurance premia Rs. 5,407, loan repayment Rs. 26,929, depreciation Rs. 14,041 and miscellaneous expenditure Rs. 2,105.

In the non-mechanised households, expenditure is very low. The expenditure has been only for travelling, transport of fish and to purchase some equipments and also to keep the remaining fish in the ice factory and so on. In the I group, the average expenditure for equipment is Rs. 1,210 and miscellaneous expenditure Rs. 7,452. Total operational expenditure is Rs. 8,662. For the II group, the average operational expenditure is Rs. 9,004. For equipments, the expenditure is Rs. 1883 and for miscellaneous expenditure is Rs. 7,171. In the III group (non-mechanised), the expenditure is low because most of them are not engagd in fishing. Expenditure for equipment is Rs. 2,213 and for miscellaneous expenses Rs. 3,021. The average operational expenditure for the III group is Rs. 5,234.

Work Contribution in Mechanised Group

Table 5.3 shows contribution of respondents, other females and male members in work pattern, average working hours and income. In mechanised households, respondents' work contribution is 23.8 per cent, other females 11.7 per cent and male members 64.5 per cent. The male members' work contribution is high when compared to female workers, because some of them are not going for fish sale. Earlier they were supposed to work but after mechanisation they have stopped working. Some of them do part-time working in fishing. The respondents fully engaged in fishing to the extent of 76.1 per cent, rest being part-time workers. Villagewise, higher proportion of respondent worker contribution is in Ullal (27.3 per cent) and lower proportion in Hoigebazar (17.4 per cent). The contribution of females other than the respondents is higher in Hoigebazar (30.4 per cent) and lower in Kudroli (6.5 per cent). Male worker contribution is higher in Bengre and Kudroli (67.9 and 67.4 per cent respectively) and lower in Hoigebazar (52.2 per cent). The average working hours are 8.2 hours for respondents, 8.7 hours for other females and 10.2 hours for male members. In Bengre, average working hours for the respondents are 7.8, 8.6 hours for other females, and 11.1 hours for males. In Kudroli, average working hours for the respondents are 8.3, for other females 7.6 hours and 10.2 hours for male members, For the III village (Ullal), the average working hours of respondents are 8.1; for other females 8.6 and for male workers 10.2 hours. In the IV village (Hoigebazar), the

Table 5.3: Work Contribution, Working Hours and Wages

Village	*% of contribution by*			*Average Working hours*			*Wages (Rs)*			*Total*
	Respon-dent	*Other females*	*Males*	*Respon-dent*	*Other females*	*Males*	*Respon-dent*	*Other females*	*Males*	*Family income*
Mechanised										
Bengre	22.7	9.4	67.9	7.8	8.6	11.1	5,928	1,294	31,937	39,159
Kudroli	26.1	6.5	67.4	8.3	7.6	10.2	8,050	1,940	39,191	49,180
Ullal	27.3	13.6	59.1	8.1	8.6	10.2	6,557	2,023	29,255	37,836
Hoigebazar	17.4	30.4	52.2	9.3	9.1	8.2	6,157	10,000	38,891	55,066
Average	23.8	11.7	64.5	8.2	8.7	10.2	6,757	2,703	34,833	44,295
Non-Mechanised										
Bengre	31.6	26.3	42.1	10.5	11.6	11.7	4,120	5,529	77,260	16,909
Kudroli	27.3	30.3	42.4	10.0	7.2	8.9	6,628	4,180	88,542	19,350
Ullal	28.1	31.3	40.6	9.3	11.1	9.1	5,298	3,163	12,959	21,420
Hoigebazar	23.5	43.2	33.3	10.3	10.8	11.7	6,134	6,089	4,109	16,332
Average	26.7	34.9	38.4	9.9	10.4	10.3	5,698	4,681	8,303	18,682
Group average	25.3	23.8	50.9	9.1	9.6	10.3	6,206	3,731	21,037	30,976

average working hours of respondents are 9.3, for other females 9.1 and 8.2 for male workers. Highest average working hours are in Hoigebazar for the mechanised household, (9.3 hours) because there are many dry fish processing workers requiring higher time for processing.

Income contribution by respondents for all the mechanised households is 15.3 per cent, 6.1 per cent for other females and 78.1 per cent by male workers. The respondents' contribution is low in Hoigebazar (11.2 per cent) and high in Ullal (17.3 per cent). Other female income contribution is low in Bengre (3.3 per cent) and high in Hoigebazar (18.2 per cent). Male income contribution is more or less high in all the villages, 81.6 per cent in Bengre, 79.7 per cent in Kudroli, 77.4 per cent in Ullal and 70.6 per cent in Hoigebazar. The male workers income contribution is higher than that of the female workers.

Income Contribution in Non-mechanised Group

In the non-mechanised households, the proportion of respondents' work contribution is 26.7 per cent, for other female workers it is 34.9 per cent and for male workers 38.4 per cent. The proportion of female workers contribution is higher than that of the male workers unlike in the mechanised households. The proportion of respondents' work contribution is higher in Bengre (31.6 per cent) and lower in Hoigebazar (23.5 per cent). Other female workers contribution is higher in Hoigebazar (43.2 per cent) and lower in Bengre (26.3 per cent), but the proportion of other female workers' contribution is higher when compared to contribution by respondents in all villages except in Bengre. The proportion of contribution of male workers is higher in Kudroli and Bengre and lower in Hoigebazar.

The average working hours in non-mechanised households is 9.9 for respondents and for other females it is 10.4 hours per day and for male workers 10.3 hours. The working hours are higher in Bengre (10.5) followed by Hoigebazar (10.3), Kudroli (10) and Ullal (9.3). Working hours of other females are higher when compared to respondent working hours. The average working hours for other females in Bengre is 11.6, in Ullal 11.1 hours and lowest at 7.2 hours in Kudroli. The male working hours are highest in Bengre and Hoigebazar (11.7) than in Ullal (9.1) and Kudroli (8.9). In the non-mechanised households, the working hours are higher as most of them are full-time workers.

In the non-mechanised households, the proportion of respondents' income contribution is 30.5 per cent for all the villages; it is 25.1 per cent for other females and 44.4 per cent for males. The proportion is higher in Hoigebazar (37.6 per cent) and lower in Ullal and Bengre (24.7 and 24.4 per cent respectively). For other females, the proportion of income contribution is higher in Hoigebazar (37.3 per cent) and lower in Ullal (14.8 per cent). The female worker income contribution is higher in Ullal (60.5 per cent) and lower in Hoigebazar (25.1 per cent). In all the four villages, female income contribution is higher except in Ullal. Here male workers' income contribution is higher because some households have professionally employed members.

Household Expenditure

Table 5.4 describes the household expenditure pattern. In the entire sample, the average household expenditure is Rs. 22,537. In the mechanised households, the average expenditure is Rs. 31,176 and in the non-mechanised households, it is Rs. 16,419. The pattern of expenditure in the mechanised households shows that, food is the major item of expenditure accounting for 50.1 per cent; fuel expenditure has a share of 3.8 per cent and clothing, health, education and other expenditure share 13.3, 5.8, 9.4, 17.6 per cent respectively.

In the non-mechanised households also, the proportion of expenditure shows that food is the major item of household expenditure accounting for 53.7 per cent. Fuel, clothing, health, education and others share 5.8, 12.9, 4.5, 6.6 and 16.5 per cent respectively. The proportion of expenditure on food and fuel is higher in the non-mechanised households, and for other items more or less same in both the types of households. But the absolute amount of expenditure is higher in the mechanised households for all the items.

Contribution in Processing and Marketing Activities

The average working hours for processing households in the mechanised group is 8.2 hours, but higher at 10.3 hours in the non-mechanised groups. For the marketing households, the average working hours are somewhat lower at 8.2 hours for mechanised and 9.7 hours for non-mechanised groups. The wages of the respondents are higher in the processing households when compared to marketing households especially in the mechanised group. The income expenditure ratio is more favourable to the processing households in both the groups.

Table 5.4: Total Expenditure of Different Households

(in Rs)

Village	*Food*	*Fuel*	*Clothing*	*Health*	*Education*	*Others*	*Total*
Mechanised							
Bengre	17,000	1,489	4,768	1,789	2,609	6,075	33,730
Kudroli	15,121	958	3,936	2,135	3,596	5,397	31,143
Ullal	12,586	918	2,858	1,542	2,356	3,500	23,760
Hoigebazar	16,095	1,166	4,300	1,300	2,666	6,300	31,827
Total	15,618 (50.1)	1,188 (3.8)	4,153 (13.3)	1,807 (5.8)	2,908 (9.4)	5,501 (17.6)	31,176 (100.0)
Non-mechanised							
Bengre	8,670	833	2,079	699	2,183	1,606	16,070
Kudroli	9,392	850	2,111	1,000	1,111	2,555	17,019
Ullal	10,448	803	2,500	755	944	3,944	19,394
Hoigebazar	7,066	1,151	1,789	558	878	1,984	13,426
Total	8,824 (53.7)	941 (5.8)	2,124 (12.9)	719 (4.5)	1,091 (6.6)	2,717 (16.5)	16,419 (100.0)

Figures in parentheses denote percentage.

Decision-making by Respondents

The important decision variables relate to work condition and work mobility, family maintenance, social functions, asset purchase, education, health, leisure, entertainment, etc. It is observed that nearly half (47.9 per cent) of the respondents point out that the decisions pertaining to work are taken exclusively by themselves. According to one-fifth (20.8 per cent), the decisions are taken jointly, that is in consultation with all the members of the family. About 22.9 per cent of the respondents are not consulted at all in the decision process. It is evident from the foregoing observation that women are involved in decision making as a great majority of them play a significant role in the matters relating to work and mobility.

The classification of data on the basis of mechanised and non-mechanised households exhibit a very important fact that the role of women in decision making is prominent in the case of non-mechanised households when contrasted to women of mechanised households. However, the participation of women in decision making process among both mechanised and non-mechanised households is deplorable except in work and mobility. There is no significant differences among the respondents regarding their role in decision making in different villages.

The role played by the respondents who are engaged in processing activity is better when compared to those who are engaged in marketing activities. Among the respondents who are engaged in processing activities from the mechanised and non-mechanised households, the latter play a better role compared to the former in all the matters except in education and health. For purchase of assets, women who are engaged in processing activity from the mechanised households do not play any major role. On the contrary, 13.3 per cent from the non-mechanised households take decisions with regard to the purchase of asset exclusively by themselves. The role of women in decision making seems better with regard to work and mobility and household maintenance compared to account keeping, purchase of assets education and health leisure and entertainment.

The analysis of data of the respondents who are engaged in marketing shows that except on work and mobility their role in decision making is limited in other aspects. About 63.3 per cent and 75 per cent of the respondents belonging to mechanised and non-mechanised households take major decisions exclusively by themselves regarding work and mobility, whereas in the case household maintenance, account keeping, purchase of assets, education, health, leisure and entertainment, their role is not prominent. Furthermore, the respondents from the non-mechanised households play a better role in decision making than in mechanised households.

Female-headed Households

In the entire sample, the proportion of female-headed fishing households is higher at 49 per cent. In the mechanised households, the proportion of FHHs is 27.1 per cent and in the non-mechanised households it is higher at 69.2 per cent. The average family size of a mechanised household is 6.92 and the average number of family workers 3.3. The respondents' education level is 30.7 per cent illiteracy, 23.1 per cent primary level and 46.2 per cent middle and high school level. Their contribution to income is Rs. 7,554 for the entire mechanised households with a total family income of Rs. 44,502. In this, the percentage of fish income is 93.1 per cent. The working hours for the respondents is 7.7 hours. Total household expenditure is Rs. 29,724 and assets including the house amount to Rs. 1,40,907.

In the non-mechanised households, the average family size is 6.83 and the average family worker size 4. The educational level of respondents is 16.7 per cent illiteracy, 19.4 per cent primary level, 55.5 per cent middle and high school level and 8.4 per cent college level. The average income of the respondents is Rs. 5,809, with a total family income of Rs. 19,216. The proportion of fish income in this is 83.4 per cent. The respondents average working hours are 9.9 hours. The average household expenditure is Rs. 16,361 and assets (including the house) amount to Rs. 56,399. Variables like income, expenditure, asset, etc, show that the values in the mechanised group are higher when compared to the non-mechanised group.

Impact of Technology

When a new technology is introduced it has many social and economic consequences. The outlook of the individuals in society undergoes a change. Attitudes towards family changes in work pattern, labour migration, imparting of skills, division of labour and entrepreneurship and communication undergo changes as a consequence of technology adoption.

Technology may have favourable impact on better health conditions and yield less hazardous work, more leisure and better education, better social participation, institutional development, etc. It may also affect the small fishermen negatively who cannot compete with mechanised fishermen and hence the gap between the two may even increase.

The impact of technology on fisherwomen is evaluated in terms of the nature of work, working hours, intensity of activity, income generation, attitude formation and decision-making. Though at the macro level, new technology in fish catch, processing and marketing is very much visible, at the individual levels except the ownership of mechanised boats, other aspects of modernisation in processing, marketing, etc, may not be quite prominent owing to the fact that these activities are not organised at the micro-level. In processing of dry fish, normally traditional methods are adopted in a large scale. At the village level however, marketing is a mix of both modern and traditional methods.

Impact of Work Conditions

The different technology levels refer to mechanised and non-mechanised boats, traditional and modern methods of processing and marketing. The level of mechanisation can be different in different village clusters. For example, Bengre and Kudroli have a larger number of samples having purse-seiners and trawlers while Ullal and Hoigebazar have a larger number of non-mechanised boats.

A higher level of mechanisation reduces not only the drudgery of work but also working hours especially of females engaged in different fisheries activities like processing and marketing. In the case of households having mechanised boats, 29.2 per cent of women are not engaged in fishing activities. This is a process of diverting their labour to other activities less hazardous than fishing. The proportion of women not engaging themselves in fishing is only 1.0 per cent in the non-mechanised cases. Processing here involves only that of dry fish processing. In the sample, marketing households are larger in number. Where there is adequate drying facilities for fish processing like in Hoigebazar, processing activity is concentrated. In the mechanised group, the proportion of processing households is only 11.8 per cent whereas it is 29.4 per cent in the case of non-mechanised group. Normally, processing would be in large scale in the mechanised group where labour hiring will also be resorted to. In the non-mechanised group, processing is normally undertaken in a small scale. During the monsoon season, no processing is being done and respondents in the mechanised group seek other types of work; in the non-mechanised group, some respondents are found to do marketing of dry fish or repairing nets, etc.

In the case of marketing households, all the members have to go to Bunder (the main landing centre) to buy fish and transport it to the selling points. In the mechanised group, all the members sell the fish in the specified market places whereas, in the non-mechanised group, 27.8 per cent of members sell fish on headloads travelling from house to house or in the market centres or on the roadsides. The marketable quantity in the mechanised cases is larger, fish being transported by trucks and stored in modern ice plants. In the non-mechanised cases, 19.4 per cent of the respondents carry fish on head from Bunder to the selling points. The quantity marketed is small and storage done in houses in traditional ways. A few mechanished households are found to be marketing fish in wholesale also.

During rainy season, the non-mechanised boats will be busy in fish catching and hence females in that group will also be working during that season mainly in marketing of fresh and dry fish. In the mechanised households however, only a small proportion of members would be undertaking fish marketing in the rainy season. In some of the Ullal non-mechanised households, the respondents are found to be making bidis during the rainy season.

In the non-mechanised households, the work contribution of respondents is higher (26.7 per cent) to total family income than in the mechanised households (23.8 per cent). Bengre, Ullal and Kudroli in the non-mechanised and Ullal and Kudroli in the mechanised groups have a larger work contribution by the respondents. When income contribution is considered, the share of the respondents is 30.5 per cent in the non-mechanised and 15.3 per cent in the mechanised groups. And relatively speaking, females in the non-mechanised households have a better productivity (share in income contribution being higer than in work contribution) when compared to the mechanised households. The productivity ratio (income contribution divided by work contribution) of the non-mechanised households works out to 1.14, whereas it is only 0.64 in the mechanised households. In Bengre and Ullal, non-mechanised households have a lower productivity and all the villages in the mechanised group have lower productivity.

In terms of different activities, the income contribution in marketing is slightly higher than that in processing in the non-mechanised cases (31 per cent as against 29.8 per cent) in the mechanised households. However, income contribution is more or less similar (15.8 per cent) in processing and in marketing (15.5 per cent) activities.

The duration of daily working hours in the on-season is higher for the non-mechanised respondents (9.9 hours) as against 8.2 hours for the mechanised households. The working hours for the marketing households are lower when compared to that in processing in both the groups. Not only for the respondents but also for other family members the working hours are more in the non-mechanised households.

Table 5.5 describes the impact of technology on different activities in different villages. In the mechanised households, all the working respondents (34 in number) are fully aware of the advantages modern technology would bring in to different activities. Of the 14 non-working respondents, four (two each in Bengre and Kudroli) are not aware of the advantage of mechanisation; 10 members are partly aware of the effects of technology. In the non-mechanised households (51 working women), only 30 can appreciate the advantage of the new technology. It is evident that female respondents who are engaged in fishing activities in the mechanised group have responded very well to mechanisation.

All the households in the mechanised group have modern methods of marketing (organised selling points), transport (trucks) and storage facilities. Among the non-mechanised households, 19.2 per cent still adhere to traditional transport methods (carrying on head) and 30.8 per cent store fish in the houses in a traditional way.

Impact on Decision-making

The attitude towards an relationship with family and other members is somewhat better in the mechanised households. Major decision-making by the female respondents in work and mobility is satisfactory in the non-mechanised households (78.9 per cent as against 47.9 per cent in the mechanised households). In households maintenance and religious functions, 38.5 per cent of respondents take major decisions in the non-mechanised households whereas it is 22.8 per cent in the mechanised households. In education and health and also in leisure and entertainment, the proportion of females taking major decisions is higher in the non-mechanised households. Only in the case of purchase

Table 5.5: Impact of Technology on Different Activities

Village	Awareness			Improved	Marketing		Transport		Storage		Life style		
	1	*2*	*3*	*Production*	*A*	*B*	*A*	*B*	*A*	*B*	*Improv-ed*	*Aver-age*	*Deterio-ration*
Mechanised H/H													
Bengre	12	5	2	19	19	-	19	-	19	-	10	6	3
Kudroli	12	2	2	16	16	-	16	-	16	-	7	6	3
Ullal	6	1	-	7	7	-	7	-	7	-	3	3	1
Hoigebazar	4	2	-	6	6	-	6	-	6	-	4	1	1
Total	34	10	4	48	48	-	48	-	48	-	24	16	8
Non-mechanised H/H													
Bengre	3	-	3	-	4	2	4	2	3	3	3	2	1
Kudroli	5	1	3	-	6	3	7	2	6	3	3	4	2
Ullal	13	4	1	-	15	3	16	2	15	3	4	9	5
Hoigebazar	9	2	8	-	17	2	18	1	12	7	10	7	2
Total (Number)	30	7	15	-	42	10	45	7	36	16	20	22	10

1 = Fully aware; 2=Partly aware; 3=Not aware.
A = Modern; B=Traditional.

of asset and account keeping that females in the mechanised households take a better decision when compared to their counterparts in the non-mechanised households.

Processing households in both the groups take a better decision in work and mobility than in marketing households. In other variables, it is normally better in the marketing households.

Impact on the FHHs

Though the household income is higher in the mechanised households in the FHHs, the respondents in the FHHs have a better income contribution. The average wages of the respondents in the FHHs is Rs. 7,554 per annum in the mechanised and Rs. 5,809 in the non-mechanised households as against Rs. 6,460 and Rs. 5,448 in the male-headed households for the two groups. The FHHs have a slightly better total income than the MHHs in both the groups—Rs. 44,502 and Rs. 19,216 as against Rs. 44,281 and Rs. 17,381 respectively.

Table 5.6 describes the impact of technology on the FHHs. The proportion of FHHs is higher in the case of non-mechanised groups. Of the 13 FHHs in the mechanised groups, all the households are aware of the beneficial effects of mechanisation. They adopt modern methods of marketing, storage and transport. Of the 36 FHHs in the non-mechanised groups, 22 are aware of the beneficial effects of technology while 10 respondents feel that they are doubtful, and four respondents are aware only some of the benefits. The households own only traditional boats or engaged in processing and marketing. In marketing, 29 households (80.6 per cent) adopt modern methods of marketing like selling in organised places. The remaining households (19.4 per cent) sell fish by carrying from house to house or on the roadsides. Thirty households hire out trucks for transport of fish while 6 carry on head from Bunder to the selling points. It may be mentioned here that these households are also practising traditional methods of marketing and also of storage. In storage, about 75 cen cent of the households make use of cold storages and 25 per cent store fish in the houses.

Regarding awareness of technology in the FHHs of the mechanised group, only 50 per cent are aware of the beneficial effects of technology. Thirteen households practise modern methods of marketing while 3 adopt traditional methods. Only one household of the total 16 adopt traditional transport method. But the proportion of households adopting traditional storage method is more than 30.6 per cent.

Table 5.6: Impact of Technology on FHHs

Village	Awareness			Production		Marketing		Transport		Storage		Life Style		
	1	2	3	A	B	A	B	A	B	A	B	Improved	Average	Deterioration
Mechanised H/H														
Bengre	6	-	-	-	6	6	-	6	-	6	-	2	3	2
Kudroli	4	-	-	-	4	4	-	4	-	4	-	3	-	1
Ullal	-	-	-	-	-	-	-	-	-	-	-	-	-	-
Hoigebazar	3	-	-	-	3	3	-	3	-	3	-	2	-	-
Total	13	-	-	-	13	13	-	13	-	13	-	7	3	3
Non-mechanised H/H														
Bengre	2	-	2	2	-	3	1	3	1	4	-	2	1	1
Kudroli	2	1	1	2	-	3	1	3	1	2	2	2	1	1
Ullal	11	3	-	11	-	12	2	14	-	13	1	5	6	3
Hoigebazar	7	-	7	7	-	11	3	10	4	8	6	5	6	3
Total (Number)	22	4	10	22	-	29	7	30	6	27	9	14	14	8

1 = Fully aware; 2=Partly aware; 3=Not aware.
A = Modern, B-Traditional.

Table 5.7: Impact of Technology on Different Age-groups

Age Group	Awareness			Improved Production	Marketing		Transport		Storage		Life Syle		
	1	2	3		A	B	A	B	A	B	Improv-ed	Aver-age	Deterio-ration
Mechanised H/H													
Less than 25 years	3	2	2	7	3	-	3	-	3	-	4	2	1
26-40 years	13	7	2	22	13	-	13	-	13	-	9	9	4
Above 41 years	18	1	-	19	18	-	18	-	18	-	11	5	5
Total	34	10	4	48	34	-	34	-	34	-	24	16	10
Non-mechanised H/H													
Less than 25 years	4	1	5	-	7	3	8	2	6	4	3	5	2
26-40 years	12	3	6	-	17	4	18	3	15	6	8	8	5
Above 41 years	16	3	4	-	18	3	19	2	15	6	9	9	3
Total (Number)	32	7	15	-	42	10	45	7	36	16	20	22	10

1 = Fully aware; 2 = Partly aware; 3 = Not aware.
A = Modern; B = Traditional.

Impact on Different Age-groups

Table 5.7 describes the impact of technology on different age-groups. Of the 34 working (respondent) households, 18 belong to above 41 years age-group, 13 in the 26-40 year age-group and only 3 in the below 25 year age-group. Of the 14 non-working households, 7 in the 26-40 year age-group are aware of some of the effects of technology. Two each in the lower age group are not aware of the beneficial effects. The older age group performs better in the non-mechanised cases also. Of the 21 households belonging to the above 41 years age-group, 14 are aware, 3 partly aware and 4 not aware of the beneficial effects. In the middle age-group, only 57 per cent are aware of the technology. The proportion is only 40 per cent in the lower age group. The proportion of households adopting modern methods of marketing, storage and transport is normally higher in the upper age-group and lower in the lower age-group suggesting a direct relationship between experience and technology awareness.

Conclusion

Of the 34 working households in the mechanised group, 24 households express better living conditions after the introduction of mechanisation both directly and indirectly, the other 6 households reporting average improvement and four expressing slight deterioration in their conditions. Of the non-working 14 respondents, 10 report average improvement while 4 have deterioration. In the FHHs, the proportion of better-off conditions is 53.8 per cent as against 50 per cent for all, suggesting only 48.6 per cent better-off cases for MHHs. In the different age-groups, the proportion of better-off cases is higher for the older age-groups (61.1 per cent) when compared to the middle (40.9 per cent) and younger (57.1 per cent) age-groups.

In the non-mechanised cases, the proportion of better-off cases is only 38.5 per cent. It may be mentioned here that because of mechanisation even the non-mechanised households would benefit indirectly by way of getting higher employment in the mechanised boats, better infrastructure for marketing, storage and transport and higher quantum of fish for sale. The proportion of households reporting deterioration is slightly higher than in mechanised households (19.2 per cent as against 16.7 per cent). In the FHHs, the proportion of better-off cases is around 38.8 per cent, suggesting a slightly lower proportion for the MHHs. The proportion of better-off cases is higher for the older age group (42.8 per cent) as against the middle (38.1 per cent) and younger (30 per cent) age groups as in the mechanised group.

Notes and References

1. Alfred D'Souza, "Women in Contemporary India", in M. Ali Khan and Aysha (ed), *The Status of Women in India*, Delhi: Manohar, 1975, p.1.
2. E.M. Rogers and E.F. Shoemaker, *Communication of Innovations: A Cross-Cultural Approach,* New York: Free Fress, 1971, p.7.
3. Mumtaz Ali Khan and Noor Aysha, *Status of Rural Women in India,* New Delhi: Uppal Publishing House, 1982, p.7.
4. Robert K. Merton, *Social Theory and Social Structure,* New Delhi: Amerind Publishers, 1968, Chap.X.
5. Also see Robert K. Merton and Alice S. Kitt, "Reference Groups" in Lewis Coser and B. Rosemberg (ed), *Sociological Theory.* Fourth Edition, New Delhi, Macmillan Publishing Co., 1976, pp. 243-250 and C.P. Loomis and Z.K. Loomis. *Modern Sociological Theories,* New York, Robert E. Krieger Publishers, 1975, Chap. V.

VI

Marine Fishermen Cooperatives in Kerala

Katar Singh

Abstract

Marine fisheries are an open access resource i.e., a resource which is nobody's property and hence is freely accessible to anybody who wants to exploit it. This has made it difficult to manage the resource because in the absence of property rights it is not possible to regulate and coordinate its use and exclude the free riders. Consequently, marine fisheries have been over-exploited in many coastal areas of India including Kerala. Co-operative management of marine fisheries seems to hold high promise as an instrument of managing them on sustainable yield basis as well as improving the socio-economic wellbeing of marine fishermen. The Government of Kerala has taken a number of pioneering measures including the establishment of marine fishermen co-ops to improve the well-being of fishermen. This chapter presents an overview of the origin, growth, and present status of marine fishermen co-ops in Kerala and examines their role and impact.

Introduction

Kerala is well known for its rich marine fish resources, especially prawns and oil sardines. It has a coastline of 590 km and a continental shelf area of about 36,000 sq. km. It ranks second in terms of total inland and marine fish production and first in terms of marine fish production in India. In 1990-91, its total fish production was estimated at 5.51 lakh mt of which 5.14 lakh mt (93%) was contributed by the marine fisheries (GOWB 1992:35-36). In 1988, Kerala contributed about 20% of the India's total marine fish catch and accounted for nearly 35% of the country's total earnings from the export of marine products. Kerala thus occupies a place of prime importance in the marine fisheries sector of India's economy.

Marine fisheries constitute an important sector of Kerala's economy as well. Nearly 3% of the state's total population is engaged in this sector. The contribution of the fisheries sector to the state's net domestic product over the period, 1980-81 to 1989-90, has ranged from 1.30% to 2.09%; in 1989-90, it was 1.54% (GOK 1991 a). As of March 1990, the total fishermen population in the state was estimated at about 9.54 lakh comprising 7.48 lakh of marine fishermen and 2.06 lakh of inland fishermen. Of the total fishermen, some 1.90 lakh (20%) were active sea going fishermen (GOK 1991 b). Marine fishermen in the state are concentrated in some 222 coastal villages. Of the 14 districts in the state, 9 districts, viz., Trivandrum, Quilon, Alleppey, Ernakulam, Trichur, Malappuram, Calicut, Kannur and Kasargod, are maritime districts

Despite the significant role that the marine fishermen play in the economy of the state, their standard of living is not satisfactory. The fishermen lag behind the other sections of the population in terms of per capita income, literacy, access to public health care and other socio-economic indicators of development. Duly recognising the need for state intervention to ameliorate their socio-economic conditions, the Government of Kerala (GOK) launched a number of welfare programmes for fishermen including organisation of marine fishermen's co-operative (co-op) societies.

In this chapter, we examine the origin, growth and development, organisation structure, impact, problems, and prospects of marine fishermen's co-ops in Kerala. This is done partly on the basis of review of literature and partly through case studies of a few selected co-ops in the state.

The Origin and Growth of Fishermen Co-operatives

Kerala has a long history of experience with fishermen co-ops. The Princely State of Travancore-Cochin had initiated the establishment of fishermen co-ops as early as 1917 and by 1933, 95 societies had been set up. The co-operatives were, however, not effective as they did not protect the fisherfolk from the depredations of merchants and middlemen (Kumar 1988). Policies favouring the artisanal fishermen continued after Independence. The GOK was perhaps the first state in India to have recognised the need for organising fishermen co-ops for improving the lot of fisherfolk and to enact a legislation, the Kerala Fishermen Welfare Societies Act 1980 that provided for constitution of

Fishermen Welfare Societies (FWS) and initiation of welfare and development programmes exclusively for fishermen.

Consequently, some 222 FWS were organised rather hurriedly all along the coast covering the entire population of active marine fishermen. The FWS so organised, however, remained more or less inactive until the Kerala State Co-operative Federation for Fisheries Development Limited popularly known as MATSYAFED was set up in late 1984 and took up the task of reorganising them into Fishermen Welfare and Development Co-operative Societies (FWDCS). Table 6.1, presents information by district about the number of fishing villages, number of FWDCS, and number of Matsya (Fish) Maveli Stores set up by the co-ops in the state.

Table 6.1: District-wise Number of Fishing Villages, Fishermen Development and Welfare Co-operative Societies and Matsya Maveli Stores in Kerala, 1990

Sl. No.	*District*	*No. of Fishing Villages*	*No. of Societies*	*No. of Stores*
1.	Trivandrum	42	18	5
2.	Quilon	27	12	7
3.	Alleppey	30	10	5
4.	Ernakulam	21	8	3
5.	Trichur	18	6	1
6.	Malappuram	23	9	2
7.	Kozhikode	34	9	5
8.	Cannanore	11	4	1
9.	Kasargod	16	5	3
	All	222	81	32

Source: Kerala Fisheries: Facts and Figures 1990, Department of Fisheries, Government of Kerala, Trivandrum, 1991, pp. 218-221, 232, and 241-43.

As shown in the table, as of 1990, there were 81 Fishermen Development and Welfare Co-operative Societies in Kerala. They run Matsya Maveli Stores for the benefit of their members. In 1990, these were some 31 such stores in the nine maritime districts of the state.

Organisation Structure of Fishermen Co-operatives

There are two apex organisations at the state level- MATSYAFED and the South Indian Federation of Fishermen Societies (SIFFS)- that are engaged in organising and assisting fishermen co-ops in the state. At the state level, a Cabinet Minister is incharge of development of fishermen and fisheries in the state. The Secretariat Department of fisheries is headed by a Secretary to the GOK and the technical Department of Fisheries by a Director. The MATSYAFED has a clear mandate of co-operativisation of traditional fishermen and spearheading of development programmes for them. It has a three-tier structure with Primary Fishermen Co-operative Societies (PFCS) at the village level, district level offices, and the state level federation. The main activities of the MATSYAFED include organisation of the PFCS, procurement of fishing equipment and inputs, establishment and operation of manufacturing facilities such as net factory and ice plants, implementation of fisheries development programmes through the PFCS and development of infrastructural facilities. As of 1992, the MATSYAFED had organised 81 PFCS. The PFCS distribute to their members production inputs and equipment such as kerosene, nets, outboard engines, and spare parts, disburse loans and collect loan repayment instalments, and arrange for sale of fish catch of their members through auctions. The MATSYAFED provides managerial subsidy to PFCS for the first five years on a tapering basis—100% in the first two years, 80% in the third year, 70% in the fourth year, and 50% in the fifth year. The main task of the district level offices is to facilitate the flow of funds and inputs to the PFCS.

The MATSYAFED is governed by a Board of Directors comprising nine members. As of April 1992, all the nine members were nominated by the GOK. The Secretary, Department of Fisheries, GOK is ex-officio Chairman and Managing Director of the MATSYAFED. The Managing Director is assisted by a General Manager and a team of six Executives, two Accounts Officers, one Assistant Manager, Personnel, one Assistant Executive Engineer and one Assistant Engineer. The MATSYAFED owns and operates one net complex, one nylon net factory, one fish meal plant and one fleet section.

At the district level, seven District Officers look after the fisheries development work in the maritime districts and at the village level Project Officers (PO) are responsible for implementation of fisheries development programmes. The POs are ex-officio Secretaries of the PFCS. The Directorate of Fisheries implements programmes of fisheries development in non-maritime districts of the state. At the district level, District Fisheries Officers assisted by field staff are responsible for implementation of such programmes.

As already mentioned, besides the MATSYAFED, there exists another organisation, SIFFS, at the state level that promotes organisation of fishermen co-op societies in Kerala and Tamil Nadu. The SIFFS is a non-governmental organisation (NGO) registered under the Travancore-Cochin Literary, Scientific and Charitable Societies Act XII of 1955. The main objectives of SIFFS are to: organise fishermen into co-operative societies, facilitate the marketing of fish collected by its member societies, assist member societies with funds in the form of loans and grants, establish and run manufacturing units for fishing craft and gear, processing units for fish and fish products, and undertake export, import and other activities in the interest of its member-societies. The area of operation of SIFFS extends to the whole of Kerala, Tamil Nadu, Andhra Pradesh and Karnataka. But as of March 1993, its activities were confined to Quilon and Trivandrum districts in Kerala and Kanyakumari district in Tamil Nadu only. The SIFFS serves as an apex organisation in the three-tier co-op structure having District Fisheries Federations at the middle (district) level and primary fishermen co-op societies at the village level. As of 1992, it had three district level federations, one for each of the three districts where it is operating, affiliated to it and 50 primary fishermen co-ops with total membership of some 6,000.

Thus, at present, these exist two parallel organisations, MATSYAFED and SIFFS, with almost a similar mandate. The MATSYAFED functions more like a government organisation although *de jure* it is a co-operative organisation. It suffers from most of the weaknesses that afflict a typical government organisation. Although the SIFFS is an apex organisation of fishermen co-ops, it itself is not a co-operative organisation. Its major strength lies in organising and educating fishermen and developing and trying new fishing craft and gear. There is no co-ordination of the activities of the MATSYAFED and the SIFFS and consequently there is a lot of unnecessary duplication of services and facilities and hence wastage of resources.

Introduction of New Fishing Technology

Ocean fishing has been an age-old occupation in the coastal villages of south-west India as well as in other coastal states of India. The indigenous fishing methods and techniques have evolved over thousands of years and are as diverse as the local sea and social-cultural conditions prevailing in the fishing villages. The first official attempt to modernise the fishery sector in the state started in 1953 when an Indo-Norwegian project came into being. The project aimed, among other things, at introducing motors for propelling artisanal craft but it did not succeed in doing so. It then shifted its emphasis to developing new designs for mechanised boats that were to be operated from harbours. Till the 1960s, there were, however, only a handful of mechanised boats in the state and almost all of the marine fish catch came from the indigenous craft propelled by wind and manpower. The early sixties saw the introduction of a few hundred gill-net boats which were mostly complementary to the artisanal craft fleet. A watershed in the history of the fishery sector in the state came in the 1970s when Penaeid prawns suddenly found a lucrative market in the world. The high export demand at high prices led to the introduction of small 32 feet-long coastal trawlers which were capable of catching the prawns in quantities required to cater to the growing demand. The state government provided further boost to trawling to promote exports of prawns. Trawling was found to be highly profitable and consequently the seventies saw a mad rush for owning trawlers. A large number of non-fishermen capitalists moved into the sector and made heavy investment in trawlers. With the new technology came in the state-sponsored domestic and foreign capital into the fisheries sector and this proved to be against the interest of poor artisanal fishermen. Fisherfolk who see the sea as the Mother Goddess found their livelihood, ecology, and culture threatened.

The growing depletion of fish stocks in coastal waters, increasing competition with mechanised trawlers, increasing fish prices, liberal import policies which allowed import of outboard engines easily, and government loans and subsidies all speeded up the pace of motorisation of artisanal boats. Consequently, the number of mechanised fishing craft in the state increased from 983 in 1980 to 9,914 in 1991 (Anonymous, 1991 a).

In a nutshell, the artisanal sub-sector characterised by low investment, low operating cost and use of renewable human and wind energy got transformed into a modern sub-sector requiring high investment, high operating costs, and non-renewable fossil fuel for its operations. The self-reliance and sustainability characteristics of the artisanal sub-sector gave way to dependency and unsustainability—dependence on external capital, foreign technology (Yamaha diesel engines and net making machines from Japan) and non-renewable fossil fuel (kerosene) and unsustainable because the present rate of exploitation cannot be sustained.

Motorisation of traditional crafts not only calls for large initial investments but also for higher operating and maintenance expenditure year after year. This has adversely affected long term viability of motorised operations, particularly towards the end of productive life of craft and gear. This factor has become a prominent issue because of over-capitalisation in the sector during the initial phase.

Impact of Mechanisation on Fish Production

Fish catch depends upon the quantum of fish stock available which is influenced by a number of factors such as rainfall, sea surface temperature, currents, etc. Besides, factors like fishing effort and fishing technology also influence fish catch. In view of this, it is very difficult, if not impossible, to isolate and measure the impact of mechanisation on fish production; the effects of climatic and management factors are inextricably mixed up.

Initially when the number of trawlers was small, the impact appeared to be positive. Due to their ability to tap demersal resources better than the artisanal units, the trawlers gave a boost to production without significantly affecting the artisanal fishermen. When the number of trawlers increased, the story was quite different. During the late 70's, the total production itself nosedived with the mechanised boats (mainly trawlers) capturing a larger share of the declining catches. From a catch of about 400,000 tonnes in 1971, the artisanal sector's catches had fallen to about 150,000 tonnes in 1980, while the mechanised sector had improved its position steadily. An analysis of the catches shows that many demersal species have declined indicating overfishing by the trawlers. The trawlers have also damaged the natural habitat of fish like corals and small reefs leading to depletion of fish resources in the coastal waters. Thus, the 1980s saw a dismal scenario with some 3,000 mechanised boats engaging a work force of 15,000 and some 30,000 country craft employing a work force of 115,000 locked in an unfair competition for a declining fish resource. With the total fishing space and the fish stock being limited; fishing by mechanised trawlers and artisanal craft became a zero-sum game; the former cornering a higher share at the expense of the latter (Anonymous 1991a).

Pushed to the wall by these developments, the artisanal fishermen reacted strongly in the early 1980's. Their response took two forms: unionisation and motorisation. Unionisation was for pressurising the Government for granting them rights in the sea vis-a-vis the operators of mechanised boats. Motorisation was resorted to for enhancing their capacity to compete with their counterparts more effectively and to reach distant waters in search of new fish resources. The political struggle of the fishermen through their unions has been a long drawn out affair and has had only limited success. The positive result has been that the Government has understood and acknowledged the unfair nature of the competition with trawlers and has taken a pioneering step in the direction of regulation of exploitation of marine fisheries through a legislation. The Kerala Marine Fishing Regulation Act 1980 provides for regulation and prohibition of use of fishing craft in contravention of any orders issued under Section 4 of the Act, and licensing and registration of fishing vessels (GOK 1983 : 49). The main objectives of the Act are to protect the interest of artisanal fishermen, to conserve fish and regulate fishing on scientific lines, and to maintain law and order in the sea. However, the implementation of the legal provisions has been lax thanks to the economic and political clout of the trawler owners. There is a need to enforce more effectively various provisions of the Act and this cannot be done unless there is a strong political will to do so at the national level.

According to a recent study, the mean annual fish catch in Kerala in the pre-motorisation period of 1969-80 was 3.65 lakh mt. In the post-mechanisation period (1981-89), the mean annual catch was 3.90 lakh mt—about 7% higher than the mean annual catch in the pre-motorisation period. If we ignore the catch in the year 1989 (6.48 lakh mt) which was an exceptionally good year so far as fish production is concerned, the mean annual catch in the post-motorisation period (1981-88) reduces to 3.57 lakh mt which is about 2% less than the mean annual catch in the pre-motorisation period. This shows that the motorisation had hardly any significant impact on total catch. However, motorisation led to increase in the share of mechanised sector in the total catch from about 24% over the period 1969-80 to about 35% over the period, 1981-88. This was but to be expected given the higher capability of mechanised trawlers for fishing in deeper and distant waters (Anonymous, 1991 a).

Socio-Economic Condition of Fishermen

Until recently there was no reliable information available about fishermen population in the state, their socio-economic condition, and the number of fishing craft and gear used by them. A census of fishermen was taken for the first time in the state only in 1972 as part of the Eleventh All India Quinquennial Livestock Census. A more comprehensive census of fishermen in the state was conducted in 1979 by the Directorate of Fisheries, Government of Kerala (GOK, 1982). We present below some of main findings of the census.

According to the 1979 census, the total population of the fisherfolk in the state was 7.79 lakh. About 20% of the marine fisherfolk in the state were actively engaged in fishing either on a full time or part time basis. The number of active fishermen per household was 1.35. About 12% of the active fishermen worked only part-time in fishing. About 5% of the fisherfolk were engaged in other activities of which casual labour accounted for 83%, agriculture for 9% and government service for 8%.

Given the arduous nature of work involved in fishing, women are rarely engaged in fishing. According to the 1979 census, only 1.7% of the fisherwomen were directly engaged in fishing work and all of them were in the inland sector. However, fisherwomen play an important role in marketing of fish. According to the 1979 census, some 11,098 fishermen and 17,165 fisherwomen were engaged in marketing of fish in the state. They mostly carried the day's catch on head and sold it in local markets and/or to neighbouring households directly.

In the 1979 census, an attempt was made to study the pattern of income distribution among the fishermen households by ascertaining their household incomes. About 97% of the fishermen had annual incomes of Rs. 3,000 or less. Taking the poverty line at an annual income of Rs. 3,600 per annum set for Sixth Plan at 1978-79 prices, we find that almost all the fishermen in the state were below the poverty line.

Literacy and educational attainments are indicators of quality of human resource. Even though, the fishing communities are generally considered economically and socially backward, they did not lag much behind in terms of education and literacy. According to the census, nearly 65% of the males and 59% of the females were literate in the state. This compared favourably with the literacy rates of general population.

It is a well known fact that the housing conditions of the fishermen are extremely poor. Most of them live in thatched huts and *kachcha* (mud) houses. According to the 1979 census, of the total houses counted (1,18,801), about 48% were small thatched huts, about 36% were *Kachcha* houses and the remaining 16% were pucca. Most of the huts and *kachcha* houses were in bad shape and required considerable repairs and replacements. The pucca houses belonged to well-to-do fishermen and were in most cases provided by the Government or other non-governmental religious and social organisations. Lack of housing sites, electricity for lighting, potable water, and poor sanitary conditions were other problems identified in the census.

More recently, a sample survey of 250 fishermen randomly selected from five villages locates in the south-west coastal region of India was jointly conducted by the Programme for Community Organisation (PCO) and the South Indian Federation of Fishermen Societies (SIFFS) (PCO & SIFFS 1991). The main objective of the survey was to study the socio-economic changes taking place after the introduction of motorised fishing boats in the region. The survey was conducted in the year 1988-89. The main findings of the survey are summed up below:

1. About 27% of the sample fishermen did not own any fishing equipment, 54% owned motorised boats, and the remaining 19% had non-motorised boats;
2. Motorisation has been encouraged by government subsidies and low interest loans;
3. There was no real change in the borrowing pattern of the sample fishermen after the establishment of fishermen co-ops and banks;
4. The levels of borrowing and indebtedness have both increased after motorisation and a lot of capital invested in indigenous craft and gear which have now become obsolete was lying unused;
5. The establishment of co-ops had a positive impact only in one of the five sample villages;
6. Only about 14% of the sample fishermen wanted their childern to remain in the fishing occupation, 77% wanted them to go for higher education, and 39% wanted them to change to other occupations; and

7. The average annual gross income from fishing of motorised boat operators and non-motorised boat operators was Rs. 28,493 and Rs. 11,477 respectively. The corresponding figures for ringseine operators and coolies/fish vendors were Rs. 20,456 and Rs. 8,674 respectively.

In 1992, a case study was conducted by the author in collaboration with two other researchers (Balakrishnan *et al.*, 1992). The case study covered two fishermen co-ops-one affiliated to the MATSYAFED and one to the SIFFS. Its main objective was to examine the business activities of the sample co-ops and to assess their impact on their members. One of the sample co-ops (Vizhingam) was located in Trivandrum district and the other (Pallithottam-Moothakkara) in Quilon district of Kerala. From each of the two sample co-ops, 20 member fishermen were selected randomly for collection of primary data about their family size, assets, average daily catch, annual income from fishing, opinions about the co-ops, etc. The main findings of the case study are summarised below:

A Profile of the Sample Fishermen Co-operatives

Table 6.2 presents a profile of the sample fishermen co-ops covered under the case study. The Vizhinjam Fishermen Village Development Society (VFVDS) was set up in 1983. It is affiliated to the Trivandrum District Fishermen Federation (TDFF) which is the apex body for fishermen village development societies in Trivandrum set up by the Programme for Community Organisation (PCO). As of June 1992, the VFVDS had 50 members of whom 40 were active sea going fishermen. By religion, all the members were Christians. It had five paid employees who took care of its routine operations including supervision of beach level auctions, disbursement of loans, maintenance of accounts and other organisational activities. The society earned Rs. 57, 548 through commission from auctions during 1991-92, of which Rs. 44,501 had been utilised to cover the establishment expenses.

The Pallithottam-Moothakkara Fishermen Development Welfare Co-operative Society (PMFDWCS) was set up in 1987. It is affiliated to the MATSYAFED. This society was set up to facilitate the implementation of the Integrated Fisheries Development Project of MATSYAFED. The area of coverage of the PMFDWCS is much larger than that of other local fishermen societies. As of June 1992, it had 542 members of whom 250 were active sea going fishermen. But for 16 members who were Hindus, all the other members were Christians. This society employed four persons who oversaw the auctions, disbursed loans, recovered loan instalment and performed other organisational activities. The annual income of the society from commission on fish auctions in 1991-92 was Rs. 59,538 and its establishment expenses amounted to Rs. 64,300.

Table 6.2: A Profile of the Sample Fishermen Co-operative Societies

Sl. No.	*Particulars*	*Vizhinjam*	*Pallithottam Moothakkara*
1.	Name	VFVDS	PMFDWCS
2.	Year of establishment	1983	1987
3.	Membership	50	542
	Hindus	—	16
	Muslims	—	—
	Christians	50	526
4.	Active members	40	250
5.	Annual Fish catch (mt)	2,762	2,873
6.	Loan Disbursed (No.)	1	61
7.	Loan Value ('000 Rs.)	2	2,817
8.	Loan recovery ('000 Rs.)	16	232
9.	No. of employees	5	4
10.	Annual establishment cost ('000 Rs.)	54	64
11.	Annual income from commission ('000 Rs.)	58	60
12.	Annual Surplus ('000 Rs.)	13	(-)15

Source: Balakrishnan *et al.* (1992).

A Profile of the Respondents

Table 6.3 presents a profile of the sample respondents. The average size of family of the respondents from Vizhinjam was 5.5. The male female ratio for the sample households was 5:6. Fifty per cent of the heads of households covered in the survey were illiterate and only 25% had studied beyond the primary level. Family income above Rs. 12,000 per annum was reported by 75% of the respondents. Sixty five per cent of the households resided in thatched houses; none of them possessed pucca houses. Some 55% of the respondents owned land, 35% were landless, and 10% were *purampokku* (squatters).

Table 6.3: A Profile of the Sample Respondents

Particulars	*% of Sample Households*	
	Vizhinjam	*Pallithottam Moothakkara*
Family size (Numbers)	5.6	5.9
Male	5.0	6.1
Female	6.2	5.7
Education of head of households		
Illiterate	50	20
Upto primary	25	45
Above primary	25	35
Family Income (annual)		
Rs. 8,000 and below	10	20
Rs. 8,001-12,000	15	30
Above 12,000	75	50
Type of house		
Thatched	65	45
Tiled	35	50
Pucca	—	5
Land ownership		
Landless	35	15
Purampokku	10	25
Landed	55	60

Source: Adapted from Balakrishnan, et al. (1992).

The respondents from Moothakkara came from families with an average size of six. The male female ratio for the sample of households was approximately 1:1. Twenty per cent of the heads of households covered in the survey were illiterate and only 35% had studied beyond the primary level. Fifty per cent of the respondents reported family income above Rs. 12,000 per annum while 20% reported that their incomes were below Rs. 8,000 per anum. About 45% of the households resided in thatched houses: only one household possessed a pucca house. Of the respondents, sixty per cent owned land, 15% were landless, and 25% were squatters.

A Profile of the Fishing Assets

Table 6.4 presents data about ownership of various types of craft and gears and average investment in them. The south Kerala coast presents great diversity in terms of sea conditions, fish species, fishing craft and gear. Differences also exist across locations in terms of dependence on marine resources, fishing skills and social processes. For example, the southern most area in which Vizhinjam falls has a very narrow continental shelf, dense population of fishermen and extensive motorisation. In contrast, the area near Quilon has a wider shelf, less dense population and fairly high level of motorisation. Differences in fishing assets have also emerged as a result of types of technology developed for specific types of fishing environment.

Table 6.4: Pattern of Ownership and Investment in Fishing Assets in the Sample Villages

Particulars	*% of Sample Respondents*					
	Vizhinjam			*Pallithottam Moothakkara*		
	Individual ownership	*Group ownership*	*Average investment (Rs.)*	*Individual ownership*	*Group ownership*	*Average investment (Rs.)*
Type of craft						
Kattamaram	15	0	8,200	—	—	—
Kettuvallam	10	5	15,066	—	—	—
Ply vallam	60	15	18,230	20	80	17,925
Type of OBM						
8 HP	55	25	18,00	25	75	23,000
9.9 and above	45	10	21,000	—	—	—
Type of Gear						
Gill nets (Med)	80	20	48,700	25	75	43,075
Boat Seine	40	5		5	20	
Gill nets (Large)	70	15		25	55	
Trammel net	80	20		10	25	
Hooks and line	80	20		25	75	

Source: Adapted from Balakrishnan *et al.* (1992).

The survey revealed that group ownership of craft was quite low in Vizhinjam. Of the 15 ply vallams possessed by the respondents, 12 fell in the category of individual ownership. There were 20 outboard motors (OBMs) under individual ownership and only seven under group ownership. Ply vallams accounted for approximately 75% of the craft operated by the respondents. The others were kattamarams and kettuvallams. High powered OBMs (9.9 hp and above) accounted for 41% of the power devices used for propelling the craft. The major types of gear used in the area were gill nets (medium and large), boat seines, trammel nets and hooks and lines. Most of the gears were owned individually. In case of 20% of the respondents, medium gill nets, trammel nets and hooks and lines were owned collectively. Boat seines were owned by 45% of the respondents.

The survey revealed that group ownership of craft, OBMs and gear was more popular in Moothakkara. About 80% of the respondents possessed ply-vallams and OBMs under group ownership. Similarly, 75% of the respondents had group ownership of medium gill nets and hook and line gear. The ply-vallum was the only type of craft used by the respondents from Moothakkara. Only one group covered in the survey had an OBM that fell in the high power category (9.9 hp and above); all the remaining were of 8 hp or below. The more popular gear among the respondents at Moothakkara were medium gill nets, hooks and lines and large gill nets; more than 75% of the respondents had these gears. Only 25% of the respondents were using boat seines.

The sea conditions at Vizhinjam and Moothakkara partly explain the differences in the types of craft used. However, another important reason, as far as the sample is concerned, is the extent to which fishermen have come to appreciate the significance of ply-vallams as an alternative to kattamarams—the intensive extension effort made by the non-MATSYAFED societies has made an important contribution in this regard.

The average investment in the fishing craft owned by the sample fishermen ranged from Rs. 8,200 to Rs. 18,230, in OBMs from Rs. 18,000 to Rs. 23,000 and in nets from Rs. 43,075 to Rs. 48,700. This shows that the new fishing technology was highly capital intensive and beyond the reach of an average fishermen.

The Impact of Fishermen Co-operatives

The primary fishermen cooperatives have a variety of roles to play in relation to artisanal fishermen. Basically, they provide credit at a lower-than-the market interest rate through which poor artisanal fishermen are able to gain access to motorisation and modern gear for their traditional craft. This is expected to enable them to significantly improve their net returns from fishing. The MATSYAFED loans are generally given on group basis to enable those poor fishermen who are not eligible, due to their poor asset base, to get bank loans. The auction system where MATSYAFED has been able to establish a nominal presence is expected to result in higher share of fishermen in the consumer's price. The co-ops also perform the role of providing basic infrastructure and services such as housing loans, consumer stores and insurance. The Project Officers of MATSYAFED who work as Secretaries of the primary co-ops perform the role of change agents as well as moderators of local conflicts. Therefore we can see a wide variety of benefits accruing to artisanal fishermen through participation in the primary co-ops.

So far no state-wide comprehensive study has been done to evaluate the impact of fishermen co-ops on the socio-economic status of their members and on the fish stocks. Quite a few small scale sample surveys and case studies have, however, been done recently on various aspects of marine fisheries and fishermen. In assessing the impact of fishermen co-ops, we draw upon some of those studies and a case study conducted by the author in collaboration with two other researchers (Balakrishnan, *et al.* 1992).

Adoption of New Technology

The fishermen co-ops have helped their members acquire motor boats and modern nets. The mechanisation of fishing has definitely led to reduction of physical effort and drudgery in day-to-day fishing operations. The reduction in strain is expected to have a positive impact on the health and quality of life of fishermen.

The new technology has also provided the artisanal fishermen access to the outer seas. Although they do not travel great distances on a regular basis, it has made possible for them to move rapidly and harvest shoals at greater distances and with greater speed. If one does not take into account the cost involved in mechanised fishing, it would appear that the impact on production and earnings should be immense.

Technology upgradation was expected to give rise to substantial additional employment for artisanal fishermen by increasing the ability of fishermen to seek out good catches during lean seasons and by enabling them to spend longer hours at sea. The case study revealed that 95% of the respondents perceived an increase in number of fishing days after the introduction of mechanisation in both Vizhinjam and Moothakkara villages. There seems to be a high level of dissatisfaction among the respondents with ragard to the quality of technical advice and repair and maintenance facilities available. Around 95% to 100% of the respondents felt that the quality of advice available was poor and that repairs were extremely costly.

Quantum of Catch and Resource Management

Given their open access nature, marine fisheries are bound to be over-exploited and eventually suffer from what Garrett Hardin (1968) calls "the tragedy of the commons". The underlying reason for the tragedy is best summed up in the conservative dictum that "everybody's property is nobody's property" (Gordon 1954 : 135). In view of this, there is a need to regulate and coordinate fishing operations through mutually agreed rules of access and exploitation. The rules so framed will need to be legitimised through a legislative act and will also require a new frame of interpersonal relations for their implementation. This would need to be followed up with other actions capable of yielding substantial value addition, cost saving or increases in productivity. If incremental private benefits from co-operative management are likely to be very small and incremental cost in terms of loss of catch substantial, then the co-operative management will not succeed. In that case, the private benefits would need to be augmented with tapering subsidies in order to induce participation in the system.

There is a good deal of controversy regarding the extent to which over-fishing is taking place on the Kerala coast. Some artisanal fishermen operating in places with high intensity of fishing feel that they now have to travel longer distances and remain at sea for longer periods of time in order to get the volume of catch they once used to get using non-motorised fishing craft. On the other hand, if one looks at the overall production from the Kerala coast, there seems to be inadequate support for the over-fishing thesis. Some fisheries experts do opine that the lateral movement of basic seed from high seas and from less utilised coastal areas would compensate for any marginal local imbalances created in most pocktes of over-fishing in a relatively short period of time. However, it is clear that over-fishing is a very important local issue in the eyes of artisanal fishermen.

The problem of declining yield is most acutely perceived in areas which were previously high yielding and supporting large numbers of traditional fishermen. Although the cause of this phenomenon is generally attributed to trawling, the adoption of by artisanal fishermen of motorisation and ring seining has played a significant role in the over-exploitation of these waters. Perhaps the crux of the problem lies in the new technology itself which is more conducive to intensive fishing in inshore waters than an economical means of access to wider fishing grounds.

The case study examined the perceptions of the sample fishermen regarding the impact of new technology on various aspects of their lives. With regard to the change in volume of catch, all the respondents from Vizhinjam felt that instead of the expected increase, there had been a decline in catch. At Moothakkara, the pattern was a little different in the sense that only 75% felt that it was decreasing; 20% of the respondents felt that their catches had increased. About 85% of the respondents at Vizhinjam and 100% from Moothakkara observed that fish availability in the inshore areas had substantially reduced in recent times leading to decreases in both individual catches and the total catch. In other words, the perception of the respondents with regard to changes or increases in catch have by and large been negative, although the studies conducted at the state level do not bear this out (Das *et al.* 1992). Perhaps, local factors like over-fishing in areas close to urban areas, viz., Trivandrum and Quilon cities, would explain this negative perception.

Promotion of co-operative management of marine fisheries calls for substantial extension education activity in order to create among the member fishermen a shared awareness of problems, action alternatives, and opportunity cost of indiscriminate fishing. This would motivate them to participate in the co-operative management process on a sustainable basis and reduce the attempts to free ride the common pool resource management system. To sustain this change, incremental benefits to fishermen would need to be ensured, at least in the short run, through a higher share in the consumer's price, better yield, and financial support for meeting their subsistence requirements.

The focus of MATSYAFED's operations seems to be on providing the fishermen enhanced and *equitable access* to marine resources through supply of fishing craft and gear at subsidised prices. Perhaps this is the most convenient starting point for initiating a marine resource management system. However, the impact of a limited intervention can only be partial, and would need to be augmented with other activities as soon as possible.

Income and Consumption

About 95% of the respondents in both Moothakkara and Vizhinjam reported that the number of fishing days had substantially improved after motorisation. This was corroborated by the figures of increased incomes reported by 90% of the respondents from Moothakkara. However, this perception is not completely shared by the respondents from Vizhinjam of whom only 35% felt that incomes had been improving.

A partial explanation for the improvement in incomes reported in the artisanal fisheries sector can be found in the rising trend of fish prices. The more important explanation can be seen in the development of market channels and beach level auctions through which fishermen are able to realise a higher proportion of the final consumer's price. This factor assumes significance with ragard to areas like Vizhinjam and Moothakkara which are located close to major towns like Trivandrum and Quilon. All the respondents in both Vizhinjam and Moothakkara reported that price of fish had gone up after the formation of the fishermen societies in their areas.

The consumption pattern of respondents did not vary sharply across the two areas covered by the survey. Although food grains formed the major part of the consumption expenditure, expenses at tea shops accounted for nearly 18% of the total annual expenditure. Another important element in the consumption expenditure reported was liquor which accounted for 5% to 10% of the total expenditure. This reinforces a commonly held perception that drinking is an important aspect of the life of a sea going fishermen.

Besides, the fishermen have had a subsistence-orientation rather than a commercial approach to fishing and fisheries. Therefore, the time perspective of artisanal fishermen was limited to the immediate present. The temporal frame has also affected consumption habits in the community; the entire daily income is often utilised immediately or in a short period after its receipt. Almost the entire amount of income is spent on food, liquor, and repayment of loans. Perhaps the hard life that they live triggers off a need for higher levels of consumption among them. This habit of maximising the present consumption is a common characteristic of the poor everywhere. Consequently, the poor fishermen does not save any thing and is always in need of money for consumption and non-consumption purposes. This forces him to borrow from private money lenders and merchants and indebtedness has become a part of the culture of the community.

A remarkable side-effect of the improvement in income and asset-ownership was the increased credit-worthiness of fishermen in the larger community. Families who used to find it difficult to purchase rice on credit for a square meal in the past can now easily get credit from local shop keepers to meet their requirements for a week or even a fortnight. Similarly, it has also now become easier for the fishermen to borrow money for meeting his working capital requirements. This has made life much easier for him than before. However, increased credit-worthiness seems to have also increased the level of indebtedness among the fishermen.

Institutional Credit and Indebtedness

For centuries the fishermen in Kerala as well as in India have been squeezed by money lenders and merchants and this has contributed to their poverty and low socio-economic status. The co-ops were established to unshackle the fishermen from the clutches of the money lenders and merchants. The major thrust of MATSYAFED's involvement and activity since its inception had been on providing craft and gear on credit so as to enable the member fishermen to have access to modern fishing technology and thereby to better fishing spots. This approach was expected to insulate beneficiary fishermen from usurious private money lenders. However, acquisition of modern craft and gear has not ensured regular income to the fishermen; their catch and hence the income from the catch still depends to a large extent on climatic and other non-controllable factors. This compels fishermen to borrow money from private money lenders and merchants for meeting their subsistence requirements during the lean season. These loans are repaid to money lenders by diverting their produce away from MATSYAFED's and SIFFS primary auctions. As a result, the presence of MATSYAFED and SIFFS in the primary market has not gained strength. MATSYAFED offficials feel that a significant presence in the retail market needs to be built before a commanding position can be established in the primary auction market.

During the 1980's, it is estimated that bank loans to the tune of over Rs. 2 crores had been channelised to the fishermen through the co-ops. The credit requirements of fishermen being very high, they inevitably resort to borrowing from local money lenders and merchants who are easily accessible to them and whose terms and conditions are very flexible and easy. The most common method of raising loans by fishermen is by agreeing to sell their catch to the money lender. Normally the principal is never repaid by the poor fishermen. The money lender takes away from 5 to 10% of the value of fish catch as his share on a permanent basis until the principal is repaid. In some highly productive areas, this method of sharing the value of catch amounts to as high an interest rate as 50-100% per annum. Most of the fishermen in Kerala are indebted to money lenders-cum-merchants in this manner and are bound to sell their catch through them.

The institutional credit is often the least important for fishermen. Securing loans from co-ops and banks is very difficult and the terms and conditions are rigid and very difficult to fulfil. Institutional credit is often extended through various government schemes which provide grants, loans and subsidies for purchase of fishing crafts and nets. There is often a mad rush to avail of the grants, loans and subsidies. It is very rare that the poorest of the fishermen get institutional credit when they actually need it most.

Marketing

The most serious weakness of the co-ops was their failure in the marketing of fish. But for a few underdeveloped areas where the co-ops had created good local markets for fish, in other areas most of the member fishermen sold only a small part of their catch though the co-ops. There have been instances of the members and auctioneers colluding to report a lower auction value, of partial sale at high seas, and the like, in order to minimise loan repayment which is always a fixed percentage of the sale proceeds. This is easy to do because the volume of catch passing through the co-ops is relatively small when compared with the size of the market.

There were charges levelled by many of the respondents against the TDFF, the SIFFS and the MATSYAFED that they have not done enough to help them realise higher price for their catch and thereby liberate them from the clutches of the money lenders. The benefits of low interest loans, subsidies, and insurance scheme have also not reached the real needy. Generally, the complaints were genuine and had concrete bases to support them. The general feeling was that while some people have got out of the poverty trap, in general most of the fishermen were finding it difficult to survive with escalating prices of consumer durables and galloping operational costs of fishing.

Effective market intervention calls for an elaborate structure, forward integration, and ability to absorb risks in fish business. Apparently, the co-ops lack all these prerequisites and hence are not able to compete with the private trade which is well entrenched in the region and has both economic and political clout. In an interview with the author, the General Manager of the MATSYAFED admitted that his organisation was not well equipped to take up fish processing, storage, and marketing including exports and that the competition with the private trade which was doing a good job in all these areas would be very stiff in case the MATSYAFED decided to enter into those areas.

Participation of Fishermen in the Management of Co-operatives

From a broader perspective, one needs to examine the extent to which artisanal fishermen participate in making decisions that affect their access to marine fish resources and hence their livelihood. From this perspective, their participation has been low and informal. There are various reasons for low participation. Some of the reasons emanate from the open access nature of marine fisheries whereas the others can be traced to the socio-cultural, political and institutional environment under which the fishermen live and operate.

The open access nature of the resource or absence of property rights in it make it impossible to regulate and coordinate its use and exclude the free riders from fishing in the interest of the fishermen community at large. The phenomenon of free riding is conspicuous particularly in central Kerala where the mudbank phenomenon locally called *chakara* takes place. During the chakara, large shoals of fish come up near the coast and fishermen from all neighbouring areas freely share the bumper catches. Similarly, during the lean season, artisanal fishermen travel to different locations along the coast in search of better catches as a matter of routine without being challenged by the local fishermen whose catches are adversely affected.

The opportunities for participation of fishermen in the co-operative management of marine fisheries are circumscribed by the defined role of the co-ops. The starting point in people's participation is membership of the co-ops. It has been observed that a large number of fishermen are not yet members of the local primary co-ops. Moreover, many of the members tend to show little interest in the activities of the co-ops. Many of the elected boards of the co-ops have been suspended as a result of which people's faith in the democratic nature of co-ops has shaken and their participation reduced. The democratic process through which participation of fishermen in apex bodies is sought to be generated also has not been implemented.

An important factor hindering the build-up of membership has been the over-politicisation of primary co-ops. Membership in the local co-ops is the key to securing subsidies and other benefits that are channelled by the government through them. Hence political parties dominating the scene at different phases have tried to keep out individuals belonging to rival parties. The supercession of elected office bearers and appointment of administrators has only increased the sense of alienation.

It would, however, be unfair to place the entire blame on the political factors alone. At this point of time, the primary co-ops are operating with a limited mandate of providing inputs, equipment, and limited marketing and financial assistance to their members. Resource management is not their concern at all although member fishermen now realise the need for it.

There exist wide differences among fishermen in terms of skill, asset holdings and access to technology. This gives rise to various conflicts which come in the way of evolving a common pool resource management perspective. Enforcement of property rights of any kind as an instrument of resource management tends to be viewed as a threat to the survival of poor fishermen who have no other productive assets or employment to depend on.

So far as the distribution of benefits from the co-ops among their members is concerned, it was uneven. Not all the members have tried to avail of the inputs and facilities provided by the co-ops for a variety of personal, institutional and economic reasons. The problem was most acute in the case of group loans to members of the co-ops. The major issues involved related to the timing of loan delivery, timeliness in repayment and attitudes of beneficiary fishermen.

The socio-cultural milieu of artisanal fishermen also contributes to the problem of non-co-operation and low participation. Historically, artisanal fishermen have had the attitude of hunter and gatherer. The nature of fishing operations made it necessary for them to cultivate the ability to operate alone in small craft or in slightly larger craft in small groups. This left little room for developing the spirit of co-operation, collaboration and sharing of resources.

When artisanal fishermen compare their life styles with those of their counterparts engaged in other occupations, they regret having been fishermen and feel the need to at least help their children get out of this occupation. As a result, large number of children now go to schools and most of them continue their education till the secondary level. But unfortunately, the unemployment situation in the state has been acute, particularly for the first generation job seekers from the fishermen community. The educated unemployed youth are good for nothing; they are not skilled in their traditional occupation and they are not competent and educated enough to compete successfully for the limited job opportunities that are available in the state. The undercurrent of frustration has led to many social problems including alienation and communal riots.

Major Issues and Options

On the basis of the overview of marine fishermen co-ops in Kerala and the findings of various research studies presented in this chapter, many issues can be identified. In the following paragraphs, we present some of the important issues and briefly spell out practicable strategies for their resolution (Nair and Singh, 1992).

1. **Organisation Structure**

   The GOK has taken many pioneering steps including enactment of rules providing for establishment of fishermen welfare societies and regulation of fishing in the sea. It also has a Minister of Cabinet rank incharge of the Fisheries portfolio. An apex organisation, the MATSYAFED, has been established by the GOK to promote establishment of marine fishermen co-operative societies and implementing various fisheries development programmes through them. Besides, an NGO, the SIFFS, is also engaged in organising fishermen co-operative societies and helping them with marketing and credit support. Whereas this kind of organisational structure at the state level seems allright, there is a need to co-ordinate the activities of the MATSYAFED and the SIFFS and avoid unnecessary duplication of efforts. There are fishermen co-ops sponsored by both MATSYAFED and SIFFS working in the same villages and creating confusion among the fishermen and unnecessarily competing between themselves.

In out opinion, SIFFS can perhaps do a better job than MATSYAFED of organising, conscientising, and educating the fishermen and should, therefore, confine its activites to those areas. The MATSYAFED needs to be given autonomy and allowed to work like a co-operative federation and not like a GOK department. It should vigorously pursue the task of reorganising the existing fishermen societies into genuinely member-controlled fishermen co-ops and eventually establish a three-tier structure like that of the Anand-Pattern dairy co-ops.

2. Marketing of Fish

The fishermen co-ops have done precious little to help their member fishermen to realise higher price for their catch; like other private commission agents they help auction their member's catch and charge a commission for that. Higher prices for fish can be assured only if forward linkages with retailers are established and the supply of fish is regulated. This is possible only if the SIFFS and the MATSYAFED are mandated to intervene in processing, marketing, and export of fish in a big way. In our opinion, at this stage they are not equipped to do this successfully vis-a-vis the private entrepreneurs and firms that are engaged in these tasks and have developed strong economic and political clout. But there is no reason why the fishermen co-ops cannot do all these things the way the Anand-Pattern dairy co-ops have done in many states including Kerala. Here again, only a strong political will and full commitment to the co-op ideology at the state level can make a difference.

3. Management of New Fishing Technology

The GOK has facilitated and speeded up the introduction of mechanised trawlers, motorised boats and modern gear by granting low interest loans and subsidies to fishermen. In many areas this has led to over-capitalisation in the sector and over-exploitation of the marine fisheries beyond the level of maximum sustainable yield. Besides many small scale artisanal fishermen have been left worse off; they could not afford to invest so much money in modern fishing craft and gear and hence were compelled to join the rank of coolies (casual labourers). It seems to us that the sector is operating at a high input-high output level with no significant increase in the net social welfare after introduction of new technology; as a matter of fact the net social welfare may have declined due to depletion of the stock of marine fish and pollution of sea water and air caused by mechanised fishing craft. There is therefore a need to examine carefully the social desirability of new technology and the rationale of government subsidies and grants to promote it.

4. Management of Fisheries Resources

Marine fisheries everywhere are an open access, i.e., nobody can be excluded from fishing and there are no rules and regulations governing the access to and exploitation of marine fisheries. Indigenous fishing techniques and practices were congenial to exploitation of marine fisheries on a sustainable yield basis. But introduction of mechanised trawlers, motor boats, and modern gear have led to over-exploitation of marine fisheries particularly in the inshore areas and disparities in the amount of fish caught by the artisanal fishermen and the trawler owners/ operators. It is estimated that, in 1990, the per capita catch of the traditional fishermen was 3.5 mt as compared to 7.9 mt for the mechanised sector (PCO ud: 1). The phenomenal rise in marine fish landings in the years, 1989 and 1990, was due both to favourable climatic conditions and imposition of partial ban on trawlers in monsoon season. Neither the MATSYAFED nor the SIFFS has any mandate to promote the management of marine fishery resources on a sustainable yield basis. Similarly at the national level also, there is no awareness about the need to manage the marine fisheries on a sustainable basis. There is therefore a need for a national policy on fisheries that provide for, inter alia, judicious exploitation and management of marine fisheries in the interest of millions of poor fishermen who have traditionally been dependent upon this natural renewable resource for their livelihood.

The GOK has been concerned about the growing depletion of marine fisheries in the state. It has appointed three expert committees, namely, D. Babu Pal Committee in 1981, A.G. Kalawar Committee in 1984, and N. Balakrishnan Nair Committee in 1988 to study the problem, review the steps taken by it, and advise it about conservation and management of the marine fisheries. These Committees have made important recommendations for the management of marine fisheries in the state. For example, the Kalawar Committee has suggested a number of measures for conservation and management of marine fisheries. They include bringing down the number of trawlers to 1,145, motorised craft to 2,690, and non-motorised craft to 20,000 and imposing a ban on purseseining (Nair 1991: 3-4). Due to various reasons including vested interests, these measures have not been effectively implemented. There is, therefore, a need to implement them forcefully.

5. **Professional Management**

There is a need for induction of professionally trained managers at the state level and district level and for training of the primary society level staff in the basics of accounting, record keeping and sales management. Although the SIFFS and the MATSYAFED have already started the process of inducting professionals, the mission and culture of both the organisations need to be re-oriented so that they could attract and retain good professionals and provide them productive and challenging work opportunities which abound in the sector. Needless to emphasise, the professionals themselves should be thoroughly familiar with the fishermen community, marine fisheries and above all they should have a firm commitment to the co-op ideology and to the service of poor fishermen. Such professionals are a rare commodity indeed but can be found if one is really interested in them. There are many organisations within and outside Kerala where one can look for such professionals.

Concluding Remarks

To conclude, we can say that the marine fishermen co-ops in the state have yet to prove their worth. The socio-economic condition of most of the marine fishermen in Kerala has not improved significantly as a result of the establishment of fishermen co-ops. The co-ops have helped their member-fishermen in the motorisation of their boats and in acquiring modern nets and other supplies. The adoption of motorisation and modern nets has helped in increasing the average catch per unit of fishing effort but the average cost of fishing has also gone up. So, on the whole, the typical fisherman now catches more fish, earns higher income from the sale of his catch, and incurs higher expenditure on fuel, repair and maintenance of his boat, motor, and nets, and payment of interest on the money borrowed. His net income has not increased significantly. But the motorisation has reduced the drudgery involved in fishing.

On the negative side, fishermen's dependence on external agencies for fuel, motors, loans, and repair and maintenance has increased. Besides, the adoption of new fishing technology has also led to further marginalisation of many poor fishermen who have now joined the rank of coolies (casual labourers) and account for some 27% of the total population of fisherfolk. In addition, some other adverse effects of motorisation include reduction in biodiversity of fish resources in terms of mix of various species and pollution of sea water.

The GOK views the co-ops as a conduit for channelising various inputs, services, and credit facilities to fishermen. So, naturally, the fishermen join the membership of the co-ops so they could avail of those inputs and services. Most of the member-fishermen do not believe that the co-ops are their own organisations established to promote their socio-economic well-being; they view them either as a government/ MATSYAFED organisation or a SIFFS agency.

The GOK has played an important role in providing basic infrastructure in fishing villages, such as schools, roads, post offices, water, electricity, sanitation, health care facilities etc. This has resulted in improved level of living in the villages.

There is no change in the borrowing pattern of the fisherfolk after the establishment of co-ops and the entry of banks in the sector. The money lender-cum-trader still plays an important role in the lives of fisherfolk. Many exploitative practices such as *udambadi* still prevail in most of the fishing villages in the state.

The co-ops face a number of organisational and management problems which need to be resolved expeditiously. At the state level, there is a need for transforming the MATSYAFED into a truly co-op organisation and for co-ordination of its activities with those of the SIFFS. Both the MATSYAFED and the SIFFS need to focus their efforts at regulating and coordinating the exploitation of marine fisheries and managing them on sustainable yield basis. The fishermen have a high stake in the marine fisheries and they will co-operate with and support any programmes that enhance their income from this natural resource. The co-op can win their co-operation and loyalty if they intervene in a big way in fish processing and marketing and assure the fishermen a higher share in the consumer's price. If the co-ops cannot do this, they are doomed to failure and no amount of state patronage and subsidies can keep them alive for long.

Acknowledgement

This chapter draws liberally from the case studies co-authored by Nair and Singh (1992) and Balakrishnan, Singh and Das (1992). The author is thankful to his co-authors, Profs. V. Mukunda Das and Suresh Balakrishnan, and Shri Ram Mohan Nair for their ideas and insights incorporated in the chapter.

REFERENCES

Anonymous (1991 a). A Census of the Artisanal Marine Fishing Fleet of Kerala (1991). Trivandrum:South Indian Federation of Fishermen Societies.

Anonymous (1991b). Motorisation of Fishing Units: Benefits and Burdens. Trivandrum: Programme for Community Organisation and South Indian Federation of Fishermen Societies.

Balakrishnan, Suresh, Singh, Katar, and Das, V. Mukunda (1992). Interventions in Common Pool Resource Management: A Case Study of Marine Fisheries in Kerala. Paper presented at the Workshop on Co-operatives for Natural Resources Management, December 7-11, 1992, Institute of Rural Management, Anand.

Das, V.M., Balakrishnan, Suresh Padmakumar, K. (1992). Impact of Integrated Fisheries Development Project Phase I: An Evaluation. Trivandrum: Centre for Management Development.

GOK (1982). Census of Fisherfolk in Kerala. Directorate of Fisheries, Government of Kerala, Trivandrum.

GOK (1983). The Acts and Ordinances of Kerala 1981. Trivandrum: The Law Department, Government of Kerala.

GOK (1991 a). Kerala Fisheries: Facts and Figures, 1990. Trivandrum: The Department of Fisheries, Government of Kerala.

GOK (1991 b). Marine Fisheries of Kerala at a Glance. Trivandrum: The Planning and Statistical Cell, Directorate of Fisheries, Government of Kerala.

Gordon, H. Scott (1954). "The Economic Theory of a Common Property Resource: The Fishery". *Journal of Political Economy.*

GOWB (1992). *MEENBARTA,* First Issue. Calcutta: Department of Fisheries, Government of West Bengal.

Hardin, Garrett (1968). "The Tragedy of the Commons". *Science*, 162: 1243-1248.

Kumar, K.G. (1988). "Organising Fisherfolk Co-operatives in Keral". *Economic and Political Weekly.*

Nair, N. Balakrishnan (1989). Report of the Expert Committee on Marine Fishery Resources Management in Kerala. Submitted to the Government of Kerala. Vikas Bhavan, Trivandrum: Expert Committee on Marine Fishery Resources Management in Kerala.

PCO & SIFFS (1991). Small Scale Fisheries on the South-West Coast of India: A Socio-economic Study of the Changes Taking Place After the Coming of Motorisation. Trivandrum: Programme for Community Organisation and South Indian Federation of Fishermen Societies.

PCO (ud). Programme for Community Organisation: Annual Report 1990-91. Trivandrum: PCO Centre.

Nair, Ram Mohan and Singh, Katar (1992). The Marine Fishermen Co-operative Societies in Kerala: An Exploratory Study. Paper presented at the Workshop on Co-operatives for Natural Resources Management, December 7-11, 1992, Institute of Rural Management, Anand.

VII

Fish Marketing Through Cooperatives--A Study of West Bengal State

Debabrata Lahiri

Abstract

About 10 per cent of the total number of the PFCSs in the country are in West Bengal. Of which, roughly 20.00 per cent are running in profit. Still the fishermen's cooperative societies produce about one third of the State's fish production. The societies are not able establish lease hold rights on the water bodies. Also they do not get any financial support yearly as loan for fish production. But other primary societies at the village level, such as, Primary Agricultural Cooperative Societies (PACSs) get loan every year from their respective Central Cooperative Banks. The assistance from the BENFISH is also partial in many aspect to the societies. Now all assistances are given to the marine PFCSs with practically no help to the inland ones. Though at present this is being done due to availability of finance from national and international bodies, still some serious thought is required to be given for the neglected inland sector. Even the Fisheries Department with various types of schemes envisaged for the fishermen have probably forgotten the inland PFCSs, as it is evident from the nature of assistance provided to them. The commercial banks also do not want to finance the inland PFCSs. The CFCSs are mostly in moribund condition. Through provision of low quality inputs at higher prices to the PFCSs they are merely surviving in the cooperative sector. The marketing of fish seems to be nearly a forgotten area in these fishermen's societies. So this research work aims in highlighting different problem areas of fish marketing, among which, the important ones are, such as, the producer-marketing linkages, interactions with the markets and internal efficiency are dealt in greater details during the course of analysis.

Introduction

Since time immemorial West Bengal was one of the major fish consuming State in the country. In the past and also at present fish is produced in inland and as well as in marine sectors. In inland fisheries, the State's position is quite significant in production and well developed in cultural practices. In plains, inland fish are produced in tanks, ponds, lakes and rivers. In hill regions, small quantity of fishes are produced naturally in springs, locally known as 'jhora'. While the major production is obtained from inland fisheries covering all the districts, marine fish production is confined along to a much shorter coast line of Bay of Bengal in two districts e.g. 24 Parganas (South) and Midnapore (Fig. 7.1 gives the coast line). This shorter coast line is about 64 Kms in fishable area. The State has 777 sq. kms. of inshore area (upto 10 fathoms), 1813 sq. kms. of offshore (between 10 to 40 fathoms depth) and a continental shelf of 17,049 sq.kms. (upto 100 fathom) for exploration.[1]

In fisheries sector, for augmenting production there has been 148.57 per cent increase in the plan expenditure from Rs.326.27 and 811.02 lakhs from 1980-81 to 1987-88. Whereas, total production of fish from both inland and marine have increased by 64.95 per cent from 370 to 505 thousand tonnes during the same period.[2] The State is also an important producer of fish seed, which is sent to other States. Fish seed production has also increased by 3 times from 1980-81 to 1987-88.

Agencies Engaged in Fisheries Development

There has been conscious effort by the State Government to develop the poor fishermen, who are generally either landless or does not possess any water body, as they mainly belong to the Scheduled Castes community. A number of institutions are working at the villages, district and State level for development of the poor fishermen. At the State level, West Bengal Fisheries Development Corporation and State Fisheries Development Corporation are the two institutions engaged in infrastructure development, establishment of modern fish breeding farms and fish production in inland and marine sectors. At the district level Fish Farmer's Development Agencies (FFDAs) have been set up for all round development of fishermen. Financial assistances are also provided to the people below the poverty line for pisciculture through Integrated Rural Development Programme (IRDP) by the District Rural Development Agencies (DRDAs) and from commercial banks. In view of the vast potential for development marine sector, development, assistance has also been provided by various national level institutions, such as, Marine Products Export Development Agency (MPEDA), National Cooperative Development Corporation (NCDC) and National Agricultural and Rural Devçlopment Bank (NABARD).

Fisheries Development Under Five Year Plans

In the State, for past 41 years, from First to Seventh Five Year Plans the Fisheries Department have spent Rs.75.48 crores only. For the Eighth Five Year Plan alone, the estimated expenditure will be Rs.111.00 crores. Apart from this plan expenditure, in next five years development assistances will also be available from the World Bank and Danish Government to the extent of Rs.90.00 and 5.00 crores respectively. The National Cooperative Development Corporation (NCDC) will also provide assistance worth Rs.32.00 crores. Thus in next five year the total expenditure for fisheries development will be 3 times more than what had been in past 41 years.[3]

Structure of Fishermen's Cooperatives

Like other sectors of the economy, there has been a systematic development of fisheries through formation of fishermen's cooperatives within the State. The societies are being formed to bring closer the willing fishermen based on honesty and integrity. The structure of cooperative societies for inland and marine fisheries varies. At the state level, for both inland and marine sector there is one apex body known as West Bengal State Fisheries Cooperative Federation (BENFISH). In inland sector, under the BENFISH, in each district, there are Fishermen's Central Cooperative Societies (CFCSs). Depending upon the area of operation in a district there are one or more CFCSs. Under the jurisdiction of each CFCS there are a number of Fishermen's Primary Cooperative Societies (PFCSs). In the State there are about 18 CFCSs and districtwise number of PFCSs for inland and marine sector separately are given in Tables 7.1 and 7.2. Details of cooperative structure and their functions for both inland and marine sectors are given in Fig. 7.2 and 7.3.

Total number of members of PFCSs are 71,850 of which only 120.3 (1.67%) are women. Of the total number of members 60.00 and 5.00 per cent are from the Scheduled Caste and Scheduled Tribes community respectively. Paid up spare capital of the PFCSs are Rs.261.00 lakhs, of

Table 7.1 : Districtwise Number of Fishermen's Primary Cooperative Societies - Inland Sector

Sl. No.	District	No. of registered societies	No. of societies in working condition
1.	Hooghly	42	15 (35.17)
2.	Murshidabad	85	32 (39.02)
3.	West Dinajpur	29	18 (62.07)
4.	Coochbehar	71	43 (60.56)
5.	Jalpaiguri	12	12 (100.00)
6.	Birbhum	30	17 (56.67)
7.	Bankura	57	20 (35.09)
8.	Purulia	40	N.A.
9.	Nadia	82	32 (39.02)
10.	Midnapore	144	N.A.
11.	Howrah	34	8 (23.53)
12.	24 Parganas (North)	61	37 (60.55)
13.	24 Parganas (South)	110	17 (14.28)
14.	Darjeeling	2	N.A.
15.	Burdwan	39	N.A.
16.	Malda	52	N.A.
17.	Calcutta	3	N.A.
	Total	902	252

N.A. Not available

Figures in parentheses indicates percentage of working societies to total number of registered societies.

Source: Assistant Registrar of Cooperative Societies, Directorate of Fisheries, Calcutta.

Table 7.2: Number of Primary Fishermen's Cooperative Societies - In Marine Sector

Sl. No.	*District*	*No. of registered societies*
1.	Midnapore	21
2.	24 Parganas (South)	30
	Total	51

Source : Assistant Registrar of Cooperative Societies, West Bengal.
Note: As PFCS are new and huge assistance in form of loan and subsidy are available so all of them are in working condition.

which the government share is Rs.210.00 lakhs (77.78%). Cumulatively, for all PFCSs total reserve fund and working capital are Rs.22.05 and 116.00 lakhs respectively. Total capital borrowed by the societies are Rs.723.00 lakhs. The societies have fixed assets worth Rs.507.00 lakhs.

Of the total number of PFCS only 21.71 per cent are running with profit. Rest 47.45 and 30.84 per cent of them are running in loss and without any profit or loss respectively. It has also been estimated that of the total amount of inland fish production in the state about one third is from the cooperative sector.[4]

Various government schemes through which the PFCS are financially helped are:

(i) Government's contribution to share capital;

(ii) Sanctioning of loan and subsidy for purchase of crafts and gears, establishment of fish stalls in markets and purchase of fisheries requisites;

(iii) For development of infrastructural facilities, community halls and making provisions for drinking water facilities.

In addition to the above schemes, in marine sector, fishing harbours are being set up with electricity, drinking water facilities, ice factory and diesel dumps for cooperatives as well as private trawlers.

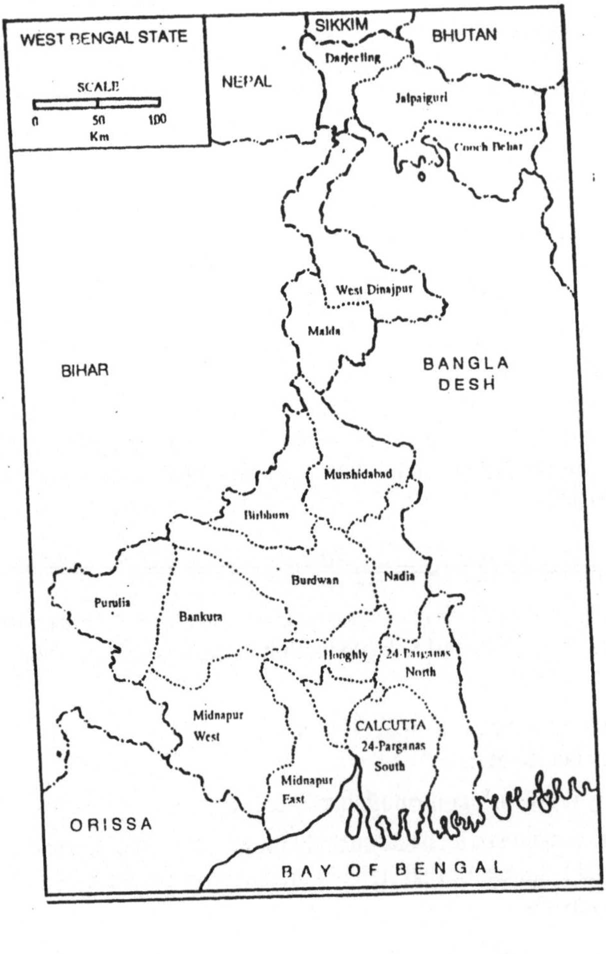

Fig. 7.1: Map of West Bengal

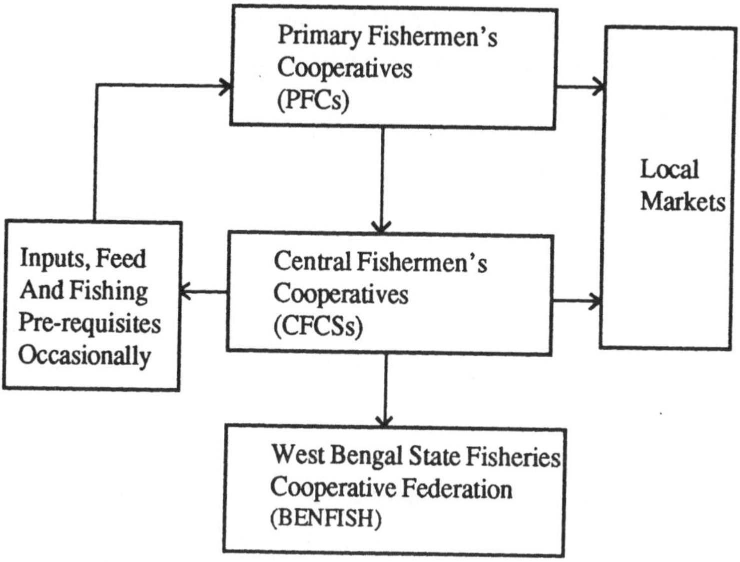

Fig. 7.2: Structure of Inland Fishermen's Cooperatives

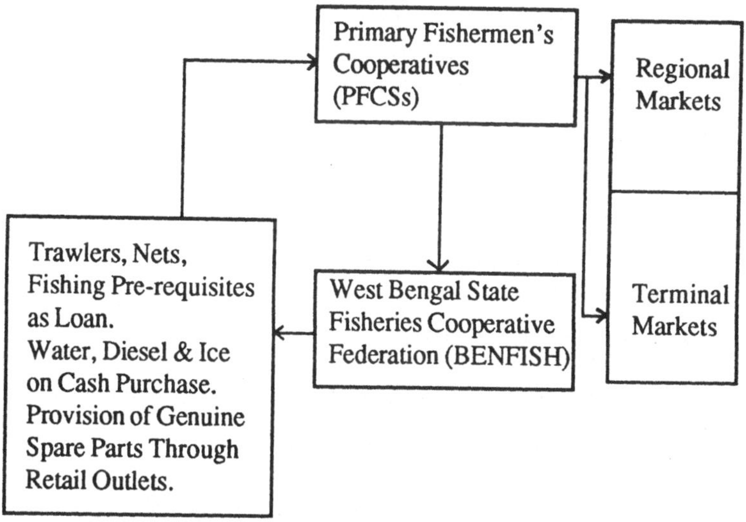

Fig. 7.3: Structure of Marine Fishermen's Cooperatives

Problems Faced by the Fishermen's Cooperatives

Among the fishermen's cooperative societies the PFCSs faces acute problems due to :

(i) As organisation of PFCSs in the State had been based on vested water bodies and along perennial rivers, so often they are found to be plagued with non-availability of vested water bodies. Even if vested water bodies are made available, they are unable to take possession due to forcible occupation by others and pending legal cases for establishment of ownership rights. The much proclaimed West Bengal Inland Fisheries Act of 1984 has so far been unable to do precious little for the PFCSs for establishing ownership rights. Even the officers of the Fisheries Department are unable to make a suitable use section 8 of this Act;

(ii) There is no budgetary provision, required for improvement of water bodies, such as, bundhs, beels and dead canals etc.

(iii) Due to absence of appropriate schemes, the riverine PFCSs faces acute problem with respect to non-availability of boats and nets;

(iv) Like other societies such as Primary Agricultural Cooperative Societies (PACSs) there is no regular flow of funds from their apex bodies in the form of loan to the PFCSs. Such absence of funds have restricted the PFCSs in their annual purchase fish seeds and feed stocking and rearing of fish. Outstanding loan to the extent of Rs.19.85 lakhs to different CFCSs and PFCSs of the State are the main reasons for absence of flow of funds.[5]

(v) Though many training programmes have been conducted for the members of PFCS for transfer of improved technology, absence of training programme on maintenance of accounts and management of societies in scientific lines has jeopardised their normal functioning. Together only 5.00 per cent of the PFCS members have received training;

(vi) Assistance by the State Government on legal aspects are also lacking;

(vii) Lack of cooperative effort among the member fishermen;

(viii) Primary weakness in forming the societies;

(ix) Financial insolvency of the societies;

(x) Restricted and partial relation between the apex bodies/societies and the primary societies.

In order to make the PFCSs viable there had been a State level conference on Fishermen's Cooperatives in February 1978. In which, the following proposals were adopted for implementation.

(i) Reorganisation of the PFCSs;

(ii) Amalgamation of moribund PFCSs to a viable one;

(iii) Declaration of bankrupt PFCSs;

(iv) To register more number of PFCSs in order to make the number of viable societies more than 400.

Fish Marketing in the State

Huge quantities of fish are required daily for consumption as about 95.00 per cent of the population consumes it.[6] In order to satisfy the internal demand, large quantities of fish are imported from adjoining States like, Bihar, Andhra Pradesh, Orissa, Uttar Pradesh, Madhya Pradesh. Also fish is imported from distant States like, Gujarat. In this State, the two largest wholesale markets are situated in Calcutta, adjacent to Howrah and Sealdah railway stations. These markets are also approachable by road. The Calcutta metropolis alone consumes more than 10.00 per cent of the country's freshwater fish production and about 30.00 per cent of the reservoir fish production, especially, carps.[7] Like fresh water fish, the State only consumes 0.83 per cent of country's total marine fish production (as fresh). Large quantities of fresh fish are consumed at other cities of the State which are situated within a distance of 200 Kms from the place of production.[8]

Calcutta city is dotted with numerous retail markets at different localities where different varieties of fish (inland and marine) are sold. These markets are not exclusively fish markets, but vegetables and groceries are also sold. Similarly, in other urban areas, municipalities and rural areas fish is sold at retail markets only. Apart from supplies to Calcutta wholesale market, supplies are also sent directly from different States to in retail markets situated in the districts.

Market Functionaries

For inland and marine fish the following types of market intermediaries are found to be performing different functions. Details of these market functionaries are given in Table 7.3. Flow of fish from producing centres to consumers takes place through a numerous combination of market intermediaries.

Table 7.3: Market Functionaries

Marine fish	*Inland fish*
Auctioneers	Wholesalers; Commission agent-cum-Wholesalers; Pre-harvest Contractors.
Commission agents (Purchase)	Commission agents (sales)
Wholesalers	Cooperative society
Commission agents (sale)	Retailer; wholesaler-cum -retailer.
Retailers	Vendor; Workers-cum-vendors
Vendors	

Prices

Prices of fish vary according to the variety, quality and its supply to the retail markets. Variation in retail prices are being discussed here according to four major varieties which are mostly available in retail markets.

(i) Carps;
(ii) Seasonal fish;
(iii) Marine fish;
(iv) Small fish.

Carps: Among major carps, such as, Rohu (*Labeo rohita*), Katla (*Catla catla*) etc. weighing upto 50 gms command the lowest price, between Rs. 20.00-24.00/Kg. Carps weiging between 1.00-1.50 kgs. command a better price around Rs.35.00 to 40.00/Kg. In the markets, a major portion of carp supply is available through local production. In Calcutta retail markets, also locally produced fish are available from suburbs.

Carps weighing around 2.50 kgs command a better price around Rs.50.00-60.00/kg. Carps weighing more than 2.50 kgs. do not command a higher price in proportionate to their weight. Also it has been observed that carps weighing more 2.50 kgs in weight are not preferred by consumers so they are available in small quantities in the retail markets. Carps weighing between 10-15 kgs. are mostly imported from outside States. Locally produced carps command a better price than imported ones. Such price differential is always between Rs.5.00-10.00/kg. Also a price differential between Rs.5.00-10.00/kg is observed between sale of whole and cut fish. In order to avoid illegal poaching and also to maximise returns the locally produced carps are sold out at lower weight.

Seasonal fish: Prices of some varieties of seasonal fish which are locally known as, Parshe (*Liza parsea*), Hilsa (*Hilsa ilisa*) and Topse (*Polynemus paradisens*) etc. vary too much depending upon their supply in religious and festive occasions. In earlier days, entire catch of Hilsa were from riverine sources of the State. But due to decrease in number of swarms of Hilsa in all the major rivers, it is now mostly obtained from marine sources. The reasons for such less number of swarms are due to construction of dams, silting of all major rivers and pollution of rivers due to discharge of industrial, agricultural and domestic effluents.

Marine fish: Only few varieties of marine fish (in fresh form) are consumed in the State. These are Pomphret (*Pompus argenteus*), Mackerel (*Scomberomorus sp.*), Tiger Prawn (*Penaeus monodon*) etc. Previously, out of these marine varieties only Tiger Prawn were consumed. But now all the other varieties are equally consumed in quantity. Mostly Pomphret and Mackerel command a better price ranging from Rs.40.00 to 48.00/kg. These two varieties are also subjected seasonally in supply. Tiger Prawns command a much higher price as bulk of the catch is processed for export. In retail markets Tiger Prawns sell between Rs.60.00-80.00/kg. During festive occasions it often goes up to more than Rs.100.00/kg.

Small fish: Among sweet water fish, certain varieties of smaller fish, such as, Koi (*Anabus testudineus*), Magur (*Clarias batrachus*), Singhi (*Heteropneustes fossilis*) etc. command a steady higher price in retail markets varying from Rs.40.00-60.00/Kg. throughout the year. Other smaller varieties like, Punti, Mourala all the year around command a higher price between Rs.40.00-60.00/Kg.

Price Stabilisation

Recently, for past few years, price of major carps in retail markets has been stabilised due to more internal production of certain varieties which grow much faster than the traditional ones and import from West Godavari district of Andhra Pradesh and Madhya Pradesh. Daily by insulated trucks large quantities of major carps come to Calcutta and to all other cities & medium towns of the State. Even this has found their place in distant markets situated in Northern parts of West Bengal, like Siliguri etc. Due to market imperfections, seasonally in production and coupled with low volume of produce the producer cooperatives are unable to influence the market price. All the factors, jointly have reduced the bargaining power of the fishermen in realisation of remunerative price. Possibly the bargaining power of the fishermen is further reduced when the commodity itself is highly perishable. The objective of cooperative marketing system as an alternative to fish marketing can increase the bargaining power of the fishermen in realisation of remunerative prices. For realisation of remunerative prices, the PFCS and CFCS will have to create strong backward linkages for increasing fish production among the fishermen and also by lowering their costs by pooling their requirements; similarly, for inputs to provide strong forward linkages by undertaking of processing of certain varieties of fish through processing of fish by-products, such as, shark liver oil etc., drying of fish on scientific lines by maintaining quality specifications and by guaranteeing access to markets and increase in bargaining power of poor fishermen's of the State. Such forward horizontal integration of functions can offer new marketing opportunities and better markets to the fishermen.

Objectives of the Study

In view of the above marketing problems of fish in the cooperatives the following are the objectives of the study:

(i) To study the producer—cooperative marketing linakges;

(ii) To study interaction with markets by the fishery cooperatives;

(iii) To study the factors affecting internal efficiency of fishermen's cooperatives.

Methodology

It has been mentioned earlier that in the State, of the 902 PFCS only 28.00 per cent are in working condition. So any attempt to study the PFCS randomly will not fulfil the objectives, as most of the societies are either moribund or defunct. Where they are in existence, they do not have much business throughout the year. In order to accomplish objectives only those PFCS are selected for study which have some reasonable amount of business. Selection of PFCS are done in consultation with district and State level officials. In inland sector 4 PFCS and in marine sector 6 PFCS have been selected and studied for this purpose. Details of the PFCS are given in Table 7.4.

Table 7.4: Name of the Sample Inland PFCSs

Sl. No.	*District Location*	*Name of the PFCSs*
1.	Midnapore	a. Mukshedpur Fishermen's Cooperative Society b. Ballabhpur Dhibar Samabay Samiti c. Dantan Sarasanka Unemployed Fishermen's Cooperative Society
2.	24 Parganas (South)	a. Captain Bheri Samabay Samiti Ltd.
3.	24 Parganas (North)	a. Chamta Fishermen's Cooperative Society
4.	Calcutta	a. Mudiali Fishermen's Cooperative Society

Findings of the Study

Findings of the study are being discussed under the following broad categories:

(i) Producer-marketing linkages;

(ii) Interaction with markets;

(iii) Internal efficiency.

While discussing the finding the main issues involved in each of the three major areas are expanded and presented below:

Producer-marketing Linakges

Producer-marketing linakges of the societies are dependent on:

(i) Assurance of better price;

(ii) Credit, inputs and services for improvement of production affecting backward linkages.

(iii) Fishermen's perception and attitude regarding coordinative functions in marketing supply of produce, supply of inputs and in providing different services.

Assurance of Better Price

Assurance for better price can be discussed here for three different channels of marketing. The first one is cooperative-consumer channel, cooperative-retailer-consumer channel and the second being cooperative-auctioneer-wholesaler (purchases)-wholesaler (sales)-retailer/vendor-consumer channel.

For assuring remunerative price of produce vis-a-vis providing fish at reasonable price to consumers the PFCSs and CFCSs have constructed retail outlets at the nearby markets. For construction of these retail outlets loans were provided to the societies. With the help of Union Ministry of Food in the Eighth Five Year Plan, the BENFISH has started an innovative method in fish marketing in Calcutta city. Fish stalls known as 'Kiosks' have been set up at identified places under the Calcutta Municipal Corporation for sale of cut and processed fish. Under this scheme, initially the BENFISH will buy fish from wholesale markets for sale through the 'Kiosks'. Initially, the BENFISH intends to utilise their knowledge and develop expertise in marketing of fish. As the scheme will make some success, the CFCSs will be asked to supply fish directly.

Inland PFCS

Cooperative-consumer channel: Problem of price for fish disposed through this channel are:

(i) In most of the societies due to small quantity of catch sales are also less. At present fish are sold only during religious and festival occasions;

(ii) Prices of fish at retail outlets situated in towns are always more than existing market prices. As a result, fish at retail outlets remain unsold;

(iii) Whereas, at the retail outlets situated in rural markets prices have been found always to be less by Rs.5.00 to 10.00/Kg. than the existing market price. Thus entire quantity of fish brought to these retail outlets are sold.

Cooperatives-Auctioneer-Wholesaler (purchase)-Wholesaler (sales)/Vendors-Consumer Channel

(i) Bulk of the fish are disposed through this channel;

(ii) The cooperatives cannot assure any steady remunerative price to the fishermen in this channel. Because they have to depend entirely on the wholesaler (purchases) and wholesaler (sales).

Cooperative-Retailer-Consumer Channel: Cooperative societies like Mudiali Fishermen's Cooperative Society sells daily about 1000 Kgs. of fish through this channel. Apart from it, carps upto 50 gms size are sold to about 30 identified retailers at fixed price. By this method the society is able to by-pass the wholesalers.

Marine PFCS

(i) For the entire fishing and trawling season spread over 6-7 months in a year quantity of fish catch is enormous. The catch also consists of a variety fish, to be consumed as fresh, certain varieties are to be processed and other varieties either small and putrified are to be sun dried. Of the sun dried varieties some are consumed and the rest which are unfit for human consumption used to manufacture poultry feed mainly;

Table 7.5: Number of Sample Marine PFCS

Sl. No.	*District/Location*	*Number of PFCS*
1.	Midnapore - Shankarpur Fishing Harbour	3
2.	24 Parganas (South) Namkhana Fishing Harbour	3
	Total	6

(ii) For marine fish, the auctioneer-cooperative-wholesaler (purchase)-wholesaler (sales)-retailers/vendors-channel is utilised solely. The reason for utilising the channel by the PFCSs is that working capital as loan is available from the auctioneers at the beginning of the fishing season.

At Namkhana, working capital loan ranging from Rs.10,000-12,000 is required for purchase of diesel, ice etc. In turn the PFCSs are morally bound to sell the entire produce through the auctioneers. The auctioneers deduct 5.00 per cent of the value of the catch in each occasion. The loan remains as outstanding throughout the fishing season.

At Digha coast, just before beginning of the season for repair of boats the PFCSs obtain loan from the auctioneer ranging from Rs. 25,000-30,000 each year. The principal is deducted gradually through sale of fish.

(iii) As a result the producer PFCS do not have any control over prices.

(iv) Instead of marketing fish by assuring better prices the BENFISH is found to be more interested in realisation of loans. Generally, about 35.00 per cent of the total value of each fish catch are realised as loan. Alternatively, also Rs.6,500.00 are required to be deposited every month by the PFCSs for recoupment of the loan.

Factors Affecting Backward Linkages Credit

At present no credit is being given to inland PFCSs by their apex bodies for renovation of tanks, purchase of fish seeds and feed etc. Even the commercial banks have been declining to advance credit. As discussed earlier, one of the reason may be due to huge amount of outstanding loan with the PFCSs. Whereas, loans have been granted to many PFCSs and CFCSs for composite duck-cum-fish culture by commercial banks. Due to various reasons the scheme has failed to make any impact on the societies. Occasionally, for fish breeding a meagre amount, as subsidised loan is available from the Fish Farmer's Development Agencies (FFDAs). Whereas, to marine PFCS subsidised credit is advanced for purchase of trawlers, nets and fishing pre-requisites. For building of boats, the members of the marine PFCSs face acute shortage of funds, as not only the release of instalments are irregular, but also due to delay in supply of wooden planks, nails and nets etc. required for construction of boats. To take the boat early for fishing at sea the members often have to borrow loans from the moneylenders.

Inputs

Inland PFCS: For inland PFCSs the main inputs which are required yearly are, mahua oil cakes, lime and fish feed. Prices of inputs provided by CFCSs to the PFCSs vary greatly as per following:

(i) PFCSs situated at distant places from their respective CFCSs have to pay 10.00-15.00 per cent more for any type of inputs than the market price prevailing at the local markets;

(ii) Whereas, the nearer PFCSs pay 10.00-15.00 per cent less than the prevailing market prices;

(iii) Differences in cost of inputs are due to the charges incurred for transporting of inputs to the distant PFCSs;

(iv) Inputs available from CFCSs are of inferior quality;

(v) Still the PFCSs have to purchase a certain portion of their required input from their CFCSs because of pressure created on them. Because, if the CFCSs are not able to sell some inputs to PFCSs it will be very difficult to keep them functioning.

Marine PFCS: For marine fishing and also for the marine PFCSs main inputs are diesel, ice and drinking water for sea going trawlers. For providing these pre-requisites a trawler whether from private or cooperative sector has to pay Rs.70.00/month for a card to the fishing harbour authorities at Shankarpur fishing harbour. At Shankarpur fishing harbour only 2.80 qtls of ice/day are sold to each trawler at the fixed price of Rs.34.00/qtl. Diesel is provided at the rate of Rs.6.40/litre and it is available in sufficient quantity. Also drinking water for sea going vessels is available in sufficient quantity. Genuine spare parts needed for repair and maintenance are also provided at fair price to cooperative trawlers through a retail shop of the BENFISH. On verbal agreement, the needed spares are provided from these retail shops on loan, to be paid within a very short time as sale of fish proceeds.

Problems faced by the marine PFCSs regarding inputs are with availability of ice at reasonable price and insufficient quantity. As per norms for a 20 qtl fish box in a trawler, almost equivalent quantity of ice is required to preserve the catch upto the fishing harbour. As the production capacity of the ice plants at the fishing harbour is low so rest of required quantity of ice is purchased from private ice plants, for which the prices are about Rs.100.00/ qtl.

Card facilities are not in existence at Namkhana because no fishing harbour has been constructed yet. Diesel, ice etc. are available freely as there are many private ice factories around Namkhana. The price of ice at cooperative ice factory fluctuates according to the market price.

Services

Inland PFCS: Services provided to inland PFCSs by their respective CFCSs are meagre and also irregular in nature. There are no defined nature of services which are to be provided regularly by the CFCSs.

Marine PFCS: But to the marine PFCSs various types of services ranging from water, electricity, auction yard, boat building & repairing facilities, ice plants, canteen and market yard are provided, which are available within the fishing harbour complex at Digha. At Namkhana, Bakhali and Raydighi, fishing harbours have not been built yet. Ice, diesel and water for trawlers are obtained from private sources only.

Fishermen's Perception and Attitude Towards Coordinative Functions

Inland PFCS: Members of the inland PFCSs feel that nature and extent of coordinative functions performed by their respective CFCSs are totally lacking. They also feel that with little effort the various functions could be performed and coordinated easily.

Marine PFCS: The members of the marine PFCSs feel that the coordinative functions performed by the BENFISH in providing credit, inputs and fishing pre-requisites are satisfactory. But for marketing of fish such functions are totally absent.

Interactions with Markets

Interaction with the markets by the PFCSs, and CFCSs and BENFISH can be discussed here for three types of markets, such as, local and external markets. Of which, within external markets there are regional, terminal, institutional and export markets. Interaction with local markets are generally done by the PFCSs directly and to some extent by the CFCSs. Interaction with external markets are possible if the CFCSs are able to pool the entire amount of produce. But the CFCSs are not able to do so due to various factors like:

(i) inadequate knowledge and fragmented selling of produce; and

(ii) quality of produce and managerial efficiencies.

Inadequate Knowledge

Inland PFCS: The inland PFCSs with their low volume of output are not able to interact fully with the local markets. But they have adequate knowledge about the prices prevailing in the local market and they take advantage of it, which is evident from the pattern and type of fish selling. To maximise returns the societies often breed fish part of which is sold out either are spawn or fingerlings. This is done for assured higher and faster returns to capital, and investment on these returns are less risky. Even some of these PFCSs have leased out a part of their own leased-in water bodies among its member fishermen for better returns. This happens only when the PFCSs are not either able to invest or manage their entire water area.

Marine PFCS: In marine PFCSs the members are engaged throughout the season for fishing and trawling. Marketing of fish is done by the respective group leaders (one group leader per trawler elected among the members of the society), who happens to be also a wholesaler (purchase) in most of the cases. Under a group leader-cum-wholesaler (purchase) another person operates, who is known as 'Challandar'. The 'Challandar' can either be an employed person or a member of the society, but he must be loyal. The 'Challandars' accompanies the fish laden truck to the terminal markets of Calcutta. The 'Challandar' even possesses much more knowledge about the markets than his group leader-cum-wholesaler (purchase). Further, it has been observed many times that these group-leader-cum-whholesaler's (purchase) price expectation has been belied due to prevailing pricing system in the Calcutta markets. Because after receipt of the consignment from the group leader-cum-wholesaler (purchase) fishes of same size and weight are further graded into two to three types. The prices of each grade of fish in Calcutta markets are also influenced by the supply of fish from other producing centres situated at other States.

Quality of Produce

Inland PFCS: The marketing arrangement of fish varies from one PFCSs to the other. For example, the Mudiali PFCS sell the entire catch of the day, according to varietywise weight of fish which are as follows:

1.	Major carps upto 50 gms. (Rohu, Catla etc.)	Rs.18.00/Kg. Sold through identified retailers only
2.	Major carps from 500-700 gms. (Rohu, Catla etc.)	Rs.22-24/Kg. Sold through identified retailers only.

3. Silver carps from 250-500 gms.	Rs.12/Kg. Sold through identified retailers only
4. Golden carps from 250-500 gms.	Rs.16/Kg. Sold through identified retailers only
5. Nylontika from 250-500 gms.	Rs.14-15/Kg. Sold through auction at the PFCS
6. Talapia	Rs.8-16/Kg. Sold through auction at the PFCS.

Marine PFCS: For marine fish quality restrictions are imposed by the auctioneers. Here also the price of fish varies according to the weight of a particular variety. Prices of export quality fish are always higher than the ordinary quality. Just after landing of the trawlers export quality fish are regularly purchased by the export houses from the sea coast itself. The extent of variation in prices of fish offered by auctioneers and export houses is given below:

Auctioneer		Export Houses
1. Hilsa	Rs.16-36/Kg.	1. Tiger prawns Rs.270/Kg. ('A' grade 50 gms & above)
2. Pomphret (below 300 gms)	Rs.76/Kg.	2. Promphret Rs.90/Kg. (above 300 gms)
3. Gurjali	Rs.25-26/Kg.	
4. Shark	Rs.5-45/Kg.	

However, some marine varieties are sold out for sun drying due to non-edible in nature and extent of purification.

Managerial Efficiency

Inland PFCSs: Managerial efficiency of the inland PFCSs is constrained mostly by absence of paid staff. There is provision of a Manager and Cashier in some bigger societies. But most of these posts either entirely or partly have remained vacant due to unsound financial position of the societies. Moreover, as payment of salaries for paid staff is to be done from the societies internal resources it becomes difficult for the societies to employ staff. Wherever paid staff has been engaged they are found to be engaged in other activities, such as, in breeding of fish rather than in financial management of the societies. Further, very few members of the societies are found to be trained in management of societies and accounts.

Marine PFCS: For marine PFCSs a Manager has been appointed for maintenance of accounts for 2-3 societies together. Salaries of Managers are entirely subsidised and paid from NCDC scheme by the BENFISH. In these societies also the number of paid staff or trained members is found to be less.

Internal Efficiency

Internal efficiency of inland and marine PFCSs is mainly influenced by the following factors:

(i) Support facilities, such as, transportation, packaging etc.;

(ii) Product differentiation;

(iii) Managerial skills in lowering cost of operations.

Transportation and Packaging

Transportation of sweet water fish to different markets by the societies has not started yet. Whereas, for marine fish of Namkhana fishing harbour, transportation by insulated trucks have been started by the BENFISH. This has not been possible at Shankarpur fishing harbour due to low volume of catch per trawler and large variation among the quantity of varieties per catch. Here, it has not been possible to pack fish individually with an identification mark of owner PFCS. Moreover, even if such packaging is done for transportation by insulated trucks, apportioning costs according to individual basket/ PFCS is found to be difficult. At Namkhana as the quantity of catch/ trawler is very near to the capacity of an insulated truck so it has been very easy to bear all the cost of transportation and packaging by an individual PFCS. Further, at this place fish is mostly obtained from deep sea fishing where a single trip varies from 7-10 days. Whereas, at Shankarpur fish is obtained both through fishing and as well as by trawling. Here the insulated trucks are found to be hired out to the wholesalers by the BENFISH for transporting of fish to different markets.

Product Differentiation

As such, no processing of fish is done by the societies for product differentiation because almost entire quantity of fish is consumed fresh at local, regional and terminal markets. As discussed earlier, product differentiation is done according to quality of fresh variety by the wholesaler (sales). A type of product differentiation is prevalent among marine fish in sun dried form. Generally, this job is done by fishermen and not by the societies.

Factors Affecting Production

A good system of marketing can be developed provided fish is produced by the PFCSs in adequate quantity and if production of different PFCSs can be pooled together at each CFCS level. Production of inland PFCSs is jeopardised due to lack of capital for purchase of fish seeds and feed. Neither the immediate apex body of PFCS, the CFCS nor the BENFISH able to provide any finance every year because these apex bodies do not receive any finance from any source for this purpose. The commercial banks are short-term based. Adoption of technology for fish production varies greatly between inland PFCSs. For example at Mudiali PFCs, where tanks are regularly stalked with fingerlings as fishing proceeds selling of fish starts right from fingerlings stage which is not more than 200 gms. By this process the society has been able to increase the production per unit of water area.

Recommendations

From the findings of the study it is evident that the fishermen's cooperative societies are weak. To overcome this situation the following steps should be initiated:

(i) Lease hold rights of the vested water bodies and stretches along the rivers are to be given only to the PFCSs on permanent basis. For conferring lease hold rights no other persons/institutions should be entertained in a village. For this Section 8 of 1984 Act must be strictly implemented by the officers;

(ii) There must be a provision of annual finance, as loan from the CFCSs to the PFCSs regularly, similar to crop loan system of the PACSs. For this BENFISH being the apex body should take active interest.

(iii) In order to maintain viability of the CFCSs, inputs and fishing requisites should not be thursted on the PFCSs. If available they should be allowed to purchase these items at a cheaper rate in the local markets only. Moreover, procurement of substandard inputs should be stopped by the CFCSs;

(iv) The BENFISH should take active interest in the development of inland PFCSs. Their activities should not be confined only for development of marine PFCSs;

(v) Sale of fish through retail stalls of PFCS should be available throughout the year to protect the consumer's interest. In urban areas for fixation of price, more understanding of the market situation is required. For this prices of fish at retail markets should changed according to the prevailing price at the markets. A little downward revision of price will not affect the profit much because in this channel (cooperative-consumer) a chain of market functionaries has already been eliminated;

(vi) Fish marketing can be initiated by the BENFISH in similar lines of the 'Cold Chain' system of Karnataka, where the apex is the sole buyer of the fish produced in cooperative sector. It is also the distributor & retailer of fish. Like Karnataka, for transporting fish there are also insulated trucks with the BENFISH. So these trucks can be utilised to a great extent in this State. If needed, few more trucks could be bought. Sale of fish through this system will enable the cooperative sector to open more retail counter in all the markets. Thus the cooperative-wholesaler (purchases)-wholesaler(sales)-retailers/vendors-consumers channel can be avoided to a great extent;

(vii) The BENFISH should start processing of marine fish in order to assure better price to the marine PFCSs for Tiger Prawns. Now it has been entirely processed in the private sector. Various forms of product differentiation of fish could be thought of at these processing units;

(viii) Production capacity of ice plants at the fishing harbours should be increased in order to cater the needs of the trawlers fully. At the fishing harbours provision of transportation of fish from the PFCSs should exist.

(ix) For improvement of knowledge, regarding management of PFCSs all the paid staff and at least all the members of the board of Directors should be trained. Such training courses should be repeated.

Notes and References

1. Government of West Bengal, Department of Fisheries (1984), *Growth of Marine Fisheries in West Bengal,* p.1
2. Government of West Bengal, Department of Fisheries (1984), *Statistical Abstract.*
3. Das, S., Asonna Matsya Biplabe Naba Naba Prajukti, Nikhut Prakshikshan - Ar. Bypak Samprasaran Karmasuchi (In Bengali), *Minbarta,* January 1992, p.9.
4. Khulbe B. and Mazumdar A. Samabay Sadasayader Jatharttia Sikshadan Byabastha Karyakan Karar Uporei Daptarer Samasta Udyog Nirvarshel (In Bengali) *Minbarta,* January, 1992, p.11.
5. The West Bengal State Fishermen's Co-operative Federation Ltd., *Annual Report,* 1991-92.
6. Gupta, V.K. *et. al. Marine Fish Marketing in India,* Indian Institute of Management, Ahmedabad, 1984.
7. Srivastava, U.K. *et. al. Fisheries Sector in India,* Oxford and IBH, P. 165, 1991.
8. Gupta, V.K. *et. al. ibid.*

VIII

Scope and Constraints of Inland Pisciculture in W. Bengal: A Case Study in Birbhum District

Kazi M.B. Rahim and Mahesh Padhy

Fisheries constitute a significant sector of the Indian economy not only from the point of view of food supplies and foreign exchange earnings but also for potential for generating employment. It contributes about 2 per cent of the gross domestic product and 3.3 per cent of the total agricultural income of the nation (CSO, 1984).

India has vast potential for fisheries, comprising of 2.02 million square kilometers (m. sq. km.) of exclusive economic zone (EEZ), 7517 km. of coast line, 29000 km. of rivers, 1.7 million hectares (m.ha.) of reservoirs, 1.383 m.ha. of brackish water area, 1.99 m.ha. of tanks and ponds area, and 1.6 m.ha. of lakes and reservoirs (Fishing Chimes, 1991).

The fish production has reached all time high i.e. from 17.56 lakh tonnes in 1970-71 to 38.36 lakh tonnes in 1990-91 an increase by 118.5 per cent. The percentage share of marine fish production to total production has declined from 61.8 per cent in 1970-71 to 59.9 per cent in 1990-91. The production of marine fish during this period has increased by 111.8 per cent. Whereas, the increase in inland fish has recorded as 129.3 per cent during the same period (Table 8.1).

With the high production of fish, the export of marine fish products rose from 33,883 tonnes in 1970-71 to 138,448 tonnes in 1990-91 (Table 8.2). India exported marine fish products worth Rs.890.4 crores

Table 8.1: Production of Fish in India

(In Lakh tonnes)

	1950-51	60-61	70-71	80-81	84-85	85-86	86-87	87-88	88-89	89-90	90-91
Marine	5.34	6.84	10.86	15.55	16.98	17.16	17.13	16.58	18.17	22.75	23
Inland	3.18	2.77	6.7	8.87	11.03	11.6	12.29	13.01	13.35	14.04	15.36
Total	8.52	9.61	17.56	24.42	28.01	28.76	29.42	29.59	31.52	36.77	38.36

Business Turnover of Fishermen's Co-operatives in India

(Rs. in crores)

	1950-51	60-61	70-71	80-81	84-85	85-86	86-87	87-88	88-89	89-90	90-91
			8.2	33.42	40.28	63.4	42.09	56.84	58.46	68	

Source: Contributed by K.L. Nalwaya, Dy. Director, N.C.U.I. in the Co-operators 1992.

Table 8.2: Export of Marine Fish Products

	1970-71	80-81	84-85	85-86	86-87	87-88	88-89	89-90	90-91
Quantity in '000 tonnes	33.9	75.6	86.2	83.7	85.8	97.8	99.8	110.8	138.4
Value Rs. in crores	35.1	234.8	384.8	398.0	460.7	531.2	598.0	635.0	890.4

Source: Contributed by K.L. Nalwaya, Dy. Director, N.C.U.I. in the Co-operators 1992.

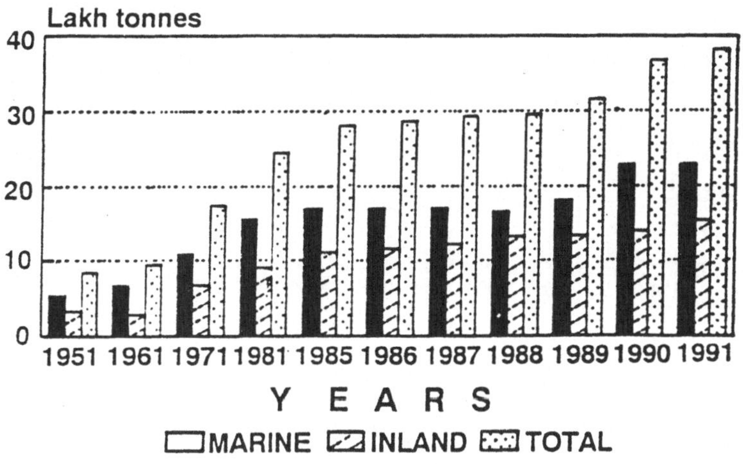

Fig. 8.1 : Production of Fish in India (In Lakh Tonnes)

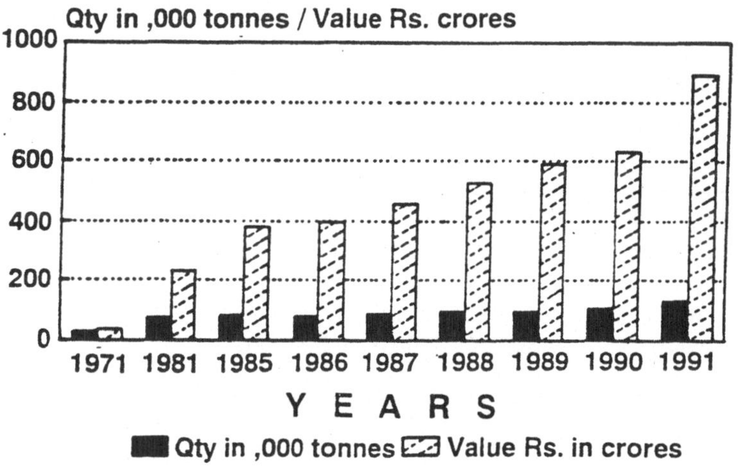

Fig. 8.2 Export of Indian Marine Fish Products

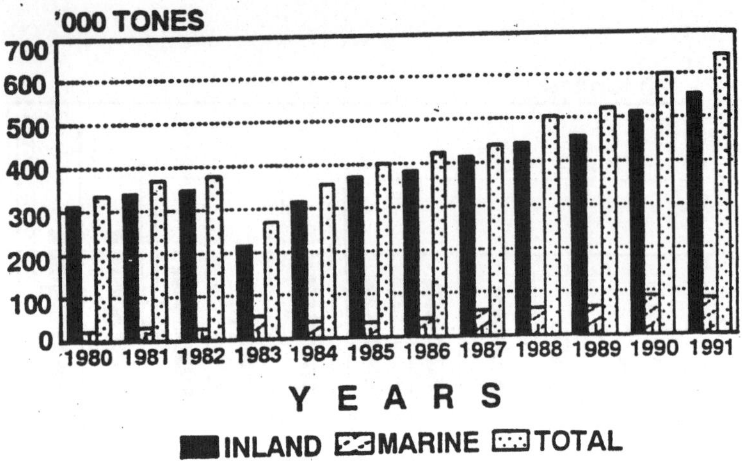

Fig. 8.3: Production of Fish in West Bengal

during the year 1990-91. India's share in the world fishery production is about 3.2 per cent. It occupies 7th or 8th position in the world fish production during last 10 years. The per capita fish consumption in India is very low, about 3.5 kg. per annum, in comparison to world's average about 12.1 kg. per annum (Fishing Chimes, 1991).

Importance and Scope of Fisheries in West Bengal

Pisciculture in West Bengal has been in vogue since long in the state. This is the only state in India where all types of capture and culture fishing are practised. Fishermen population in the state is 4,04,000, out of which 3,19,000 are inland fishermen and 85,000 are marine fishermen (1981 census). 42.5 per cent of the fishermen are literate. Fishing crafts are owned by only 30 per cent of fishermen and 61 per cent of the fishermen own fishing net. The average income per fishermen household is Rs.5300.00 per annum.

Fresh water pond and tanks constituted her most important sector of pisciculture. Out of 2.76 lakh ha. area under ponds and tanks, 70.28 per cent i.e. 2.22 lakh ha. are used for pisciculture. 19.97 per cent of the area which remains unused due to siltation and other causes could be brought under pisciculture after having proper reclamation. Besides, other potential fisheries resources of the state are beel and baor (41,782 ha.), river (1,72,587 ha.), canal (80,086 ha.) reservoir (16,739 ha.), sewage-fed-fisheries (4,083 ha.) and brackish water bheri fisheries (2,10,100 ha.) (Table 8.3).

Table 8.3: Inland Fisheries Resources of West Bengal

Fishery resources	*Total resources*	*Presently used resources*
1. Ponds and Tanks	2,76,202 ha.	2.22 lakh ha.
2. Beel and Baor	41,782 ha.	0.21 lakh ha.
3. Reservoir	16,739 ha.	0.03 lakh ha.
4. River	1,72,587 ha.	
5. Canal	80,086 ha.	
6. Sewage-fed Fishery resources	4,083 ha.	0.04 lakh ha.
7. Brackish water	2,10,100 ha.	0.34 lakh ha.

Source: Collected from the Office of WBSFCF, Calcutta.

Table 8.4 : Production of Fish in West Bengal

(thousand tonnes)

Years	*Inland*	*Marine*	*Total*
1979-80	315	22	337
1980-81	340	30	370
1981-82	350	28	378
1982-83	218	53	271
1983-84	319	36	355
1984-85	370	32	402
1985-86	384	40	424
1986-87	412	58	470
1987-88	443	62	505
1988-89	458	65	523
1989-90	512	89	601
1990-91 (target)	557	85	642

Source: Endeavour and Achievements, Dept. of Fisheries, GOWB, 1986.

West Bengal has a substantial marine fisheries resources viz. coastal line (64 sq. km.), coastal area upto the depth of 10 fathoms (770 sq. km.) high sea area having 10 to 40 fathoms depth (1813 sq. km.) and international boundary line upto 100 fathoms depth (17049 sq. km.).

The State occupies the top rank in inland fish production in India. Her share in the domestic inland fish production was 22.40 per cent in 1964 and it rose to 36.52 per cent in 1989-90. The state's contribution in the nation's total marine fish production was 3.91 per cent in the year 1989-90. The fisheries has proved itself as a very highly profitable and emerging sector of the state's. economy. The government of West Bengal has taken several measures to promote both inland and marine fisheries in the state. These measures include establishment of fishermen's co-operative and Fish Farmers Development Agencies, launching of Inland Fisheries Project and Integrated Fisheries Projects, training of fish farmers (fishermen and fisherwomen), group insurance scheme for fishermen, construction of ideal villages for fishermen project, the passage of West Bengal Inland Fisheries Act, 1984 and the enactment of the West Bengal Inland Fisheries Rules, 1985.

Till the year 1989-90, about 1,40,725 fishermen and 4,620 fisherwomen have been trained in West Bengal. The state plays an important role in fulfilling the demand of fish fingerlings in the country. About 75 per cent of the total demand of the country is met by the state (Table 8.5). Similarly, 124 fishermen engaged in marine fisheries were trained in 1980-81 and at the end of 1989-90 a total number of 2074 fishermen (marine) were trained (Table 8.6). The state has introduced a programme on social fisheries with an area of 199.30 ha. in the year 1987-88. The area under Social Fishery Programme has increased to 1890 ha. in 1989-90.

Besides, fisheries in the sewage-fed water has brought a new dimension in this field. About 4000 ha. area in the outskirt of Calcutta is being used for sewage-fed fisheries.

Brackish water fisheries, in the state, has got a great scope. The Govt. of West Bengal (GOWB) has taken several measures to increase the production under brackish water fisheries and also to improve the socio-economic condition of the fishermen engaged in this type of fisheries in collaboration with UNDP, MPEDA, and BFDA.

Table 8.5: Production of Fish Fingerlings in West Bengal

(in crores)

Years	*Quantity*
1980-81	2300
1981-82	3200
1982-83	3150
1983-84	3910
1984-85	4200
1985-86	5000
1986-87	6100
1987-88	7005
1988-89	7250
1989-90	7400
1990-91 (target)	7500

Source: Collected from Dept. of Fisheries, GOWB.

Pisciculture in the hilly watershed has also got due attention since 1986. About 450 watersheds in the hilly areas of the state have been brought under fisheries development programme in the year 1989-90.

GOWB has formulated project for pisciculture in the Kangsabati reservoir. The GOWB is also planning for spawning in different rivers of the state to increase fish production.

The use of motor boats in the marine fisheries is becoming popular in the state. During the period from 1982-83 to 1989-90 about 1,414 motorised boats have been distributed under different schemes and 7,286 fishermen farmers have been benefited in the state (Table 8.7).

Table 8.6: Training Programme of Fish Farmers over Different Years in West Bengal

Year	Training Given to					
	Inland				Masine	
	Fishermen		Fisherwomen		No. of Fishermen	Cumulative total
	No.	Camulative total	No.	Cumulative total		
1980-81	1190	1190	200	200	124	124
1981-82	1278	2468	200	400	111	235
1982-83	4728	7196	200	600	120	355
1983-84	7896	15092	200	800	160	515
1984-85	22180	37272	200	1000	180	695
1985-86	23085	60357	200	1200	259	954
1986-87	22033	82390	400	1600	280	1234
1987-88	21212	103602	1000	2600	280	1514
1988-89	20957	124559	1000	3600	280	1794
1989-90	16166	14725	1020	4620	280	2074

Source: Collected from Dept. of Fisheries, GOWB.

Table 8.7: Distribution of Boats under Different Schemes in West Bengal

Financial Assistance Provided by	Year	Motorised Boats	Non-Mechanised Boats	Benefited Fishermen
NABARD	1982-83	44	36	560
	1983-84	138	234	2506
	1984-85	202	96	2309
	1985-86	313	1703	417
	1986-87	8	85	574
	1987-88	25	34	170
NCDC	1988-89	25	-	250
	1989-90	50	-	500

Source : Collected from Dept. of Fisheries, GOWB.

Fisheries in Birbhum District

Modern pisciculture is one of the emerging activities in the district of Birbhum. It has a high potential for development of inland fisheries, particularly fresh water culture fisheries. The scope of capture fisheries is limited to only Mayurakshi and Ajoy rivers. In 1989-90, an estimated area of 17,000 ha under ponds and tanks in the district was covered under fish culture and the annual production of fish seed amounted to 60.6 crore fingerlings. According to the district Fishery Officer, Birbhum, the estimated annual production of table fish in the district in 1988-89 was about 29,000 tonnes (Singh and Bhattacharya, 1991).

The Case Study

An indepth case study was undertaken in Bolpur - Sriniketan Block of the district of Birbhum for proper understanding regarding the constraints on adoption of improved pisciculture technology. A sample of 41 fishermen (by practice) was selected randomly. The primary information required for the study were collected through personal interview method using a pre-tested survey schedule considering 1992-93 as the reference year.

Categories of Fishermen and Their Characteristics

The respondent fishermen were divided into three categories as follows:

Category I: The fishermen who produced fish for home consumption only, Category II: The fishermen who produced fish both for home consumption as well as for sale, and Category III: fishermen who performed pisciculture for business purpose only.

Out of 41 sample fishermen, 11 persons (26.83 per cent) in category I, 25 persons (60.98 per cent) in category II and 5 persons (12.19 per cent) in category III, it was noted (Table 8.8) that only one respondent fishermen by practice (2.44 per cent) belonged to original fisherman family by caste. 11 fishermen (26.83 per cent) fell in S.C. and S.T. caste and remaining 29 persons (70.73 per cent) in general Hindu caste.

Fishery appeared as the primary occupation in case of only 4.88 per cent of the respondent fishermen, secondary occupation in case of 65.86 per cent of the fishermen and tertiary occupation in case of 29.26 percent of the fisherman ((Table 8.9). It was mainly primary and secondary occupation for the fishermen conducting pisciculture for business purpose. While for other two categories of the fishermen, pisciculture appeared to be mainly secondary and tertiary occupation. The fishermen of category III had highest annual income i.e. Rs. 26,400 from the sources other than pisciculture. For the fishermen of category I and category II, these figures were Rs. 14148 and Rs. 24,560 respectively.

Regarding education it was noted (Table 8.10) that 36.34 per cent of the fishermen of category I had education upto primary level and a similar per cent of the fishermen educated above High School (H.S.) level. Only 9.09 per cent fishermen were illiterate in this category. Only 4 per cent of the fishermen in category II were illiterate, 72 per cent had education upto H.S. level, 12 per cent upto primary level and 12 per cent above H.S. In category III all the fishermen were literate. 80 per cent of them educated upto H.S. and 20 per cent over H.S. It clearly indicates that fisherman of category III do not belong to fishermen caste and are educated.

Table 8.8: Family Size and Caste of the Respondents

Category	*Family Size*	*Fisherman*	*SC*	*ST*	*General*
Category I	7.8	-	2 (18.18)	9 (81.82)	-
Category II	6.92	1 (4)	0	-	24 (96)
Category III	5.6	-	-	-	5 (100)

Operated Water Area

The culture fishery in the tanks is the main component of inland fisheries in W. Bengal. The fishermen generally possess the operatorship of ponds of different sizes, of which some are owned by them, and some are taken on lease. It is evident from Table 8.11 that the fisherman of category I operated only 0.47 bighas (3 bighas = 1 acre) of water area. Out of which 97.13 per cent area was owned by them and only 2.87 per cent area had been leased in for operation. Whereas the fishermen of category II operated a larger water area (2.723 bighas), of which 57.44 per cent was owned by them and 42.56 per cent of the water area was taken in lease. The fishermen of category III owned only 1.19 per cent of their operated water area and 98.81 per cent was taken in lease. The fisherman of this category who were engaged in pisciculture exclusively for business purposes operated largest water area among all the three categories.

Table 8.9: Importance of Fishery as an Occupation

	Primary	*Secondary*	*Tertiary*	*Average income other than fisheries*
Category I n=11	-	7 (63.64)	4 (36.36)	14418
Category II n=25	1 (4.00)	16 (64.00)	8 (32)	24560
Category III n=5	1 (20)	4 (80)	-	26400
Overall	2 (4.88)	27 (65.86)	12 (29.26)	22063

Table 8.10: Educational Status of the Respondent

	Illiterate	*Upto IV*	*Upto H.S.*	*Above*
Category I n=11	1 (9.09)	4 (36.34)	2 (18.19)	4 (36.36)
Category II n=25	1 (4.00)	3 (12)	18 (72)	3 (12)
Category III n=5	-	-	4 (80)	1 (20)

Table 8.11: Distribution of Water Area Among Different Categories of Fish Farmers

(in Bighas)

Category	*Owned area*	*Leased in area*	*Total operated land*
Category I	0.46 (97.13)	0.01.36 (2.87)	0.47 (100)
Category II	1.56 (57.44)	1.14 (42.56)	2.70 (100)
Category III	0.2 (1.19)	16.5 (98.87)	16.7 (100)

1 Acre = 3 Bighas.

Adoption and Level of Improved Pisciculture Technology

Table 8.12 depicts the levels of modern pisciculture technology over different categories of fishermen. It was noted that the application of different inputs recommended for improved inland pisciculture was highest in case of category III, followed by category II and category I.

Table 8.12: Level of Technology in Pisciculture over Different Categories

	Cat-I	*Cat-II*	*Cat-III*
Application of mahua Cake (kg per bigha)	-	8.07	15.56
Application of lime (kg. per bigha)	21.27	26.44	43.71
Application of organic manuse (Kg. per bigha)	116.05	119.43	143.71
Application of inorganic manure (kg. per bigha)	-	0.07	76.91
Fish feed (kg. per bigha)	2.90	24.23	82.63
Spawning (Rs. per bigha)	534.65	421.62	639.5?

Regarding adoption of modern technologies the fishermen of category III was best adopter. A 2-point scale (yes=1, no=0) was used to measure the adoption of technologies by different of fisherman.

It was found (Table 8.13) that the fisherman of category III adopted 58.33 per cent of improved technologies which was followed by category II (46 percent) and category I (19.69 per cent). Among the different components of modern technology, 100 per cent adoption was noted in case of mixed fish culture and lime application, under category III. Whereas 80 per cent adoption was noted in case of use of organic fertilizer, oilcake other than mahua, use of medicine and cleaning of water. Under category II 96 per cent adoption was observed in case of

Table 8.13: Adoption of Technologies by Different Categories of Farmers

Technologies	*Category I* n=11	*Category II* n=25	*Category III* n=5
Mixed Fish Culture	8 (72.73)	24 (96)	5 (100)
Chemical Fertilizer	0 (0.0)	1 (4)	2 (40)
Use of Organic Fertilizer	4 (36.30)	9 (76)	4 (80)
Lime application	7 (63.63)	24 (96)	5 (100)
Oil Cake	1 (9.09)	18 (72)	4 (80)
Mahua Cake	0 (0.0)	7 (28)	2 (40)
Cleaning of pond	2 (18.18)	11 (44)	3 (60)
Use of Medicine	1 (9.09)	17 (68)	4 (80)
Application of weed control		0.0	-
Cleaning of water	1 (9.09)	16 (64)	4 (80)
Induced breading	-	-	1 (20)
Nylon net			1 (20)
Overall adoption in %	(19.69)	(46)	(58.33)

Figures in the parentheses indicate percentage.

Mixed fish culture and lime application. 76 and 72 per cent adoption was in case of use of organic fertilizers and oil cake other than mahua respectively. Adoption of different components of modern pisciculture technology was very discouraging under category I. The adoption o mixed fish culture was measured to be 72.73 per cent and the figure wa only 63.3 and 36.36 per cent for lime application and use of organic fertilizers. For other inputs the adoption level is very negligible under this category of fishermen.

Distribution of Fishermen Among Different Adoptor Categories

Using adoption index (AI) i.e.

$$AI = \frac{\text{No. of Technologies adopted}}{\text{Total No. of Technologies}} \times 100,$$

The fisherman of different categories were distributed among different adoptor categories. The fishermen having 'O' as AI value were considered as non-adoptors, "1-33.33" as low-adoptor, "33.34-66.66" as medium adoptor and "66.67 to 100" as high adoptor. Thus it was observed (Table 8.14) that 9.09 per cent of the fishermen of category I were non-adoptors and 90.91 per cent were low adoptors. In case of category II, 36 per cent of the fishermen were low adoptors and 64 per cent were medium adoptors. 20 per cent of the fishermen in category III were high adoptors, while 60 per cent were medium adoptors and 20 per cent were low adoptors.

Table 8.14: Distribution of Adoptor Categories Among Different Categories of Farmers

Respondents	Non Adoptor	Low Adoptor	Medium Adoptor	High Adoptor
Category I n=11	1 (9.09)	10 (90.91)	-	
Category II n=25	-	9 (36.00)	16 (64.00)	
Category III n=5	-	1 (20.00)	3 (60.00)	1 (20.00

Figures in parentheses indicate percentage.

Farm Level Constraints as Perceived by Fishermen

Based on literature and prior knowledge from the survey of the area, it was hypothesised that the following factors might appear as constraints in pisciculture in the area.

1. Environmental and Situational
   (i) flood and
   (ii) drought

2. Technological
   (i) quality carps,
   (ii) disease

3. Socio-economic and Infrastructural including management-
   (i) availability of inputs including feed,
   (ii) credit,
   (iii) transportation,
   (iv) cost of production,
   (v) management,
   (vi) trading,
   (vii) extension services,
   (viii) marketing, and
   (ix) storage.

To measure the constraints for pisciculture a 4 (four) point scale was used and the score given to these points were 3 to 0 in descending order. It was noted (Table 8.15) that the constraints in the group of Socio-economic and infrastructural including management were most important among all the three groups. This group had 55.37 per cent coverage which was followed by technological group of constraints having coverage of 23.97 per cent, and environmental and situational group with only 20.66 per cent.

Considering it was imperative to ascertain the influence of these constraints individually on productivity of pisciculture, an attempt was made to establish a functional relationship between them. It was found from the fitted multiple regression equation that out of fourteen constraints considered as independent variables only five appeared to be statistically significant constraints in productivity of pisciculture.

The results may be summarised as follows:

Variables	*Regression Coefficient*	*Level of Significance*
Drought	-.31253	5 P.C.
Management	-.43494	10 P.C.
Training	-.20564	5 P.C.
Extension Service	- .14563	10 P.C.
Marketing	-.04331	10 P..C.
	$R^2 = .59826$	

Table 8.15: Relative Importance of Different Types of Constraints Among the Respondents

Constraints	*Percentage coverage of Constraints*			
	Category I n=11	*Category II n=25*	*Category III n=5*	*Over all n=41*
1. Environmental and Situational	26 (17.22)	56 (20.14)	18 (32.73)	100 (20.66)
2. Technological	40 (26.49)	62 (22.30)	19 (25.45)	116 (23.97)
3. Socio-Economic and Infrastructural including management	85 (56.29)	160 (57.55)	23 (41.82)	268 (55.37)
Total	151 (100)	278 (100)	55 (100)	484 (100)

Figures in parentheses indicate percentage.

Conclusions

1. West Bengal and the District of Birbhum have vast fisheries resources.
2. Culture fisheries is the main component of inland fisheries in the state as well as in the study area. Several efforts have been made by both government and non-government agencies to develop culture fisheries.
3. Fishermen (by practice) who performed pisciculture for business purposes were better off with respect to resource endowment (both quality and quantity).
4. Different components of pisciculture technology had been adopted by the fishermen not in toto of the recommended packages and also which was not uniform over different categories of fishermen.
5. Fishermen of category III were the high adoptors in comparison to other categories.
6. The constraints in the group of socio-economic and infrastructural including management were the most important among all the three groups of constraints.
7. Drought, management, training, extension service and marketing were among the 14 considered constraints having statistically significant influence on productivity of pisciculture.

REFERENCES

Annual Report - 1990-91, The West Bengal State Fishermen's Co-operative Federation Ltd., Calcutta.

Fishing Chimes, 1991.

Rahim MB Kazi and Singh Katar (1992), Central Fishermen's Cooperative Society, Contai--A Case Study, Institute of Agriculture, Visva-Bharati and Institute of Rural Management, Anand.

Rahim MB Kazi and Singh Katar (1992), The Marine Fishermen's Cooperative Societies in West Bengal : An Exploratory Study, Paper presented in the Symposium on Management of Rural Cooperatives, December 7-11, Institute of Rural Management, Anand.

Singh Katar and Bhattacharya, S. (1991), The Bergram Majhipara Common Pool Fish Pond : A Case Study - No. 8, Institute of Rural Management, Anand.

IX
Prospective of Fisheries in Rural Development

C.S. Singh & Malobica Das

Freshwater aquaculture resources in India comprise of small and large seasonal and perennial ponds and tanks, beals, jheels, lakes, canals, reservoirs, rivers, swamps, paddy fields and a few brackishwater bheris, which have been recently added in the freshwater community. In the past water resources in the rural areas had harboured abundant wild fish. These wild fish were used to meet the requirement of the people and due to this people showed little interest in fish culture. Even then fish farming had been a traditional practice dating back to several centuries, when the country was ruled by Hindu kings. The rulers used to excavate ponds and tanks for drinking, bathing and sometimes for small scale irrigation purposes. These ponds and tanks were used for fish keeping aimed for recreation and games rather than for commercial purposes. Tanks were also found in the temple boundaries and fish from natural resources were caught and reared in those temple tanks. People coming to the temple used to feed the fish and fish were caught and auctioned periodically after 4-5 years. The money earned was used for the welfare of the temple owners.

History narrates that fish culture in India was on a limited scale since 300 B.C. Attempt on scientific fish culture was first made in Madras some where in 1911 and culture on scientific lines was practised in Bengal during the same period. In the 20th century fish culture practices were extended to other parts of the country through establishment of Fisheries Departments in the states. It was then established that the success in aquaculture can be achieved by good breed, quality seed and nutritive feed as green revolution in agriculture was achieved by quality seed, suitable fertilizer and proper irrigation.

The carp polyculture system based on six ecologically compatible species viz. Catla, Rohu, Mrigal, Silver carp, Grass carp and Common carp, was developed by the Central Inland Fisheries Research Institute (ICAR) in sixties. This technology is well known as composite fish culture and was extended in different parts of the country with a production range of 3-10 t/ha/yr. The technology being fully tested in different parts of the country, World Bank came forward to assist Govt. of India in establishing Fish Farmers Development Agencies in different States of the country. The State Govt. then extended their full support to the fish farmers through FFDA's by means of transfer of technology for pond reclamation or excavation, providing quality seed and also by assisting in getting public tanks leased.

To meet the large requirement of quality carp seed, carp hatcheries were established under Fisheries Corporations in number of States. The carp seed production technology was also transferred to the fish farmers so as to establish their own hatcheries and meet the demand of seed to considerable extent. In order to provide scientific knowledge to the trainers, transferring technology to the fish farmers, Fisheries Research and Training Centres were established. These extension and development activities undertaken by the State Govts. popularised carp culture activities throughout the country and the fish production had gone up to the level of 3 t/ha/yr.

The important facets of successful fish culture include selection of site, proper designing and construction of ponds, selection of fish species, judicious stocking of fish, seed management and maintenance of water quality, nutritive food and feeding etc. The traditional fish farming is still being adopted by most of the farmers in the country. Though many advances have been made in aquaculture, an investment based fish farming industry could not be developed. It is obvious that we need to evolve a satisfactory, feasible, viable, adoptable and successful technology which can ensure a sustained high fish production in the ponds.

Efficient fish farm management entails special preparation of ponds to receive fish seed to new environment. For the proper, designing of fish pond/farm one should have information regarding species, stage, population and life cycles of fish for which pond is to be made. Depending upon this the depth, volume, size etc. have to be decided. For operational convenience, grow-out ponds, meant for raising table size fish should be rectangular, preferably having length: breadth as 3:1, if possible. Small ponds are preferable as they facilitate effective control of environment. Drying up of such water bodies in summer months help mineralization, removal of excess organic matter and automatic destruction of predators, minnows etc., that usually abound in perennial ponds. Keeping these in view a calendar of operation is proposed (subject to change in different agroclimatic zones of India).

Calendar of Operation

1. April, May, June
   - (i) Selection of site.
   - (ii) Soil analysis.
   - (iii) Pond construction and reclamation.
   - (iv) Stock pond and Brood pond preparation.
   - (v) Liming and fertilization.
   - (vi) Brood care.
2. July, August, September
   - (i) Brood care
   - (ii) Fish breeding
   - (iii) Fish seed rearing and stocking.
   - (iv) Sampling and maintenance of natural fish food organisms.
   - (v) Supplementary feeding.
   - (vi) Study of fish growth.
3. October, November, December
   - (i) Maintenance of fish growth.
   - (ii) Supplementary feeding.
   - (iii) Periodical control of weeds, insects and predators etc. if any.
   - (iv) Repairs of bunds, if any.
4. January, February, March
   - (i) Common carp breeding and rearing of seed.
   - (ii) Fish harvesting and marketing.
   - (iii) Annual repair of ponds.
   - (iv) Brood fish selection and segregation.

• Site Selection

Fish culture operations start with site selection. Basic requirement of a satisfactory site is as follows-

(i) The site should be non-rocky, having 2-2.5 m deep soil cover preferably near forest cover area.

(ii) Slope of the terrain gently sloping so that drainable tanks can be excavated.

(iii) Soil fractions should be 90% of the whole soil and stone and gravel not to exceed 10%.

(iv) Soil pH should be nearly neutral, total available nitrogen more than 0.1% total available phosphate 0.01%, total available carbon 1.0% and the soil should be responsive to organic and inorganic fertilizers.

(v) Sub-soil water table should not be far below deepest pond bottom during summer months in case soil is not completely water retentive.

(vi) A dependable source of perennial water should be available to fulfil the water requirement of pond.

(vii) Rate of seepage should not exceed 1 m/year.

Chemical characteristics of water greatly influence fish production in ponds. From the view-point of fish culture pH range of 6.5-8.5 is suitable for pond water. Dissolved oxygen should be 5-6 mg/1 or more, free carbon-dioxide 15-20 mg/1 and total alkalinity 100-200 mg/1. The concentration of nitrate and phosphate in water for optimum fish production are 0.2-0.5 ppm and 0.2 mg/1 respectively.

Components of Fish Farm

The essential components of fish farm are as follows—

1. A hatchery proper comprising a complex of facilities for fish spawning (spawning pool, breeding hapa etc.), hatching pool indoor & outdoor hatchery, hatching hapa) and care of hatching to raise them upto post larval stage.
2. Nursery ponds for nursing post larvae to fry stage.
3. Rearing ponds for growing fry to fingerlings.
4. Stock ponds for production of fish for marketing and to supply brood fish and donor fish for pituitary glands.

5. Brood fish ponds for rearing spawners (male, female fish to be reared separately, if possible) and also to rear spent fish. Rearing ponds can also be used as brood ponds after harvesting fingerlings.

Now the first question arises as to what should be the ratio of different types of ponds in a typical fish farm. Keeping fish industry in view a fish culturist has the option to go either to have a fish seed production farm or to produce table size fish for marketing. In case he desires to have a fish seed production farm, for this the following steps are to be considered for construction of the farm.

Let us assume that the requirement of fish seed of the area is 50 million fry or 25 million fingerling/year. Fish farmer has to consider the following points of construct the fish seed production farm.

1. Rate of stocking of post larvae is 10 million/ha.
2. Rate of survival of post larvae to fry is 50%.
3. Water surface required for 50 million fry is 10 ha.
4. The rate of stocking of fry for fingerling is 1.6 million/ha.
5. The rate of survival of fry to raise them to fingerling stage is 50%.
6. Water surface required for 25 million fingerlings is 30 ha.

Now to derive pond area for production of 100 t fish/year, in other words, to make a fish farm for fish production, the following points are to be taken care of-

1. The rate of stocking of fingerling is 6000/ha.
2. The rate of survival of fingerlings to table size fish is 90%.
3. The rate of stocking of fry to table size fish is 10,000/ha.
4. The rate of survival of fry to table size fish is 60%.
5. The average weight of marketable fish is 800 g.
6. The total fish production per year is 5 t/ha.
7. Water surface required for 100 t fish production is 22 ha.

No. of fingerlings required for an unit area can be calculated by the following formula-

$$X = \frac{P}{S \times W}$$

Where X = No. of fingerlings required
P = Production/ha
S = Rate of survival
W = Average weight at harvesting

Stocking Combination and Species Ratio

A combination of six species viz. Catla, Rohu, Mrigal, Silver carp, Grass carp and Common carp fulfills the criteria of species selection and has been proved ideal for freshwater carp culture in India. Four species (Catla, Rohu, Mrigal and Common carp) combination is also equally productive because Indian major carps fetch higher market price the Silver carp or Grass carp & Common carp. In case of non-availability of seed of exotic carps traditional three species culture (Catla, Rohu and Mrigal) can be profitably adopted.

Species ratio generally depends upon the availability of seed, nutrient status of pond and other factors hence, general guidelines can only be given which are shown in the Table 9.1.

Table 9.1: Species Stocking Ratio in Composite Fish Culture
(Stocking rate 5000 fingerlings/ha)

Species	*3 species culture*	*4 species culture*	*6 species culture*
Catla	2000 (40%)	1500-2000 (30-40%)	500-700 (10-15%)
Silver carp	-	-	1000-1500 (20-30%)
Rohu	1500 (30%)	1000-1500 (20-30%)	750-1000 (15-20%)
Mrigal	1500 (30%)	750-1000 (15-20%)	750-1000 (15-20%)
Common carp	-	1000-1250 (20-25%)	1000-1250 (20-25%)
Grass carp	-	-	250 750 (5-15%)

Selecting and stocking suitable fish species in proper ratio fulfil the primary requirement for success in fish farming. To achieve the target the second step is proper feeding of fish.

Feeding regime of fish is based on both natural and supplemental feed. Detritus, bacteria, phytoplankton, zooplankton etc. together provide natural food for fish. The natural productivity of pond is enhanced by use of manure and fertilizers which make up or provide essentially needed nutrients or minerals required for phytoplankton, serving either directly or indirectly through involved ecosystem as food for fish. The various groups of organisms follow each other in a definite sequence in a fertilized pond. They have a definite life span and therefore, exhibit periodicity or cycles of abundance. In a semi-intensive system these natural food organisms form about 50% of the total food (feeding regime). Therefore, their population must be safeguarded and promoted. The total amount of organic manure and nitrogenous and phosphatic fertilizers required for maintaining sufficient amount of plankton in the pond is based on the existing fertility of the ponds. Monthly doses of manure and fertilizers can be calculated from Table 9.2.

Table 9.2: Amount of Manure and Fertilizers Required (kg/ha/yr)

Manure/Fertilizer	*Pond type*		
	Low Productive	*Medium Productive*	*High Productive*
1. Cow-dung	6000-65000	48000-58000	25000-30000
2. Urea	225-290	156-225	150-160
or			
Ammonium sulphate	-	-	450-550
Calcium Amm. Nitrate	500-650	350-500	-
3. Single super phosphate	315-405	219-315	225-350

A supply of supplementary feed almost always results in improved production over extensive methods. It has been experienced that through systematic supplementary feeding the production of pond can be doubled or tripled. As the name suggests supplementary feeds do not provide a complete balanced diet, alone, instead they merely add extra nutrients to those already available i.e. rice bran, oil cakes are used in semi intensive culture in India where choice of materials depend upon cost and availability as well. A number of compounded feeds have been formulated, using animal protein sources i.e. fish meal, fish flour, blood meal, silkworm pupae, noto-nectids, prawn waste and very good performance is observed. However, due to high cost and unavailability of these sources, they have not been adopted by Indian fish farmers.

In recent years much attention has been focused on nutritional potential of alternative protein sources that are cheap and easily obtainable. Biogas plant slurry has also been found to be a good fish feed substitute (30-40%) oil cake soaked in slurry overnight and in the morning rice polish bran is added & made into a dough of the fish. Feed should be administered during specified hours and at fixed locations. It is preferably done in morning hours. It has how been found that feeding several times a day is more favourable in carp ponds.

In old practice feed, in dry powder form, was spread all over the pond surface. It was uneconomic as significant amount of feed goes waste. A modified method is tray feeding where thick dough of feed in the form of balls is given in shallow earthen pots which are kept at particular spots in the ponds. These spots are marked with bamboo poles fixed into pond bottom. It is important to establish a number of feeding places in the pond since fish must be fed uniformly to prevent irregular growth and loss of feed. Vegetable matter i.e. aquatic weeds, leaves and grasses for feeding Grass carp are either given as chopped pieces mixed with feed or in floating bamboo rectangular frames.

A very important consideration is the feeding schedule. Fish should be fed according to body weight and water temperature. If the pond is rich enough, abounding insufficient natural food, then supplementary feed given to the fish per day is @ 2-3% of the body weight (Table 9.3).

Table 9.3: Feeding Schedule for Table Size fish Production with A Stocking density of 5000 Fingerlings/ha

Months	*3 species culture*	*4 species culture*	*6 species culture*
First 3 months (kg/day)	2.0	2.5	3.0
Second 3 months (kg/day)	4.0	5.0	6.0
Third 3 months (kg/day)	6.0	7.5	9.0
Last 3 months (kg/day)	8.0	10.0	12.0
Total feed required (kg/yr)	1,600	2,000	2,500

Feeding spots should be periodically changed so that no putrefaction occurs. Rate and amount of feed given is reduced when temperature goes down and during peak winter it may be completely stopped. Supplementary feeding should also be stopped when thick algal blooms develop in the pond.

With a little care and by adopting the above scientific technology fish farmers can raise the production to minimum 5000 kg/ha/annum and can make a better living. There are some identified constraints also viz. technical, financial, marketing and unexpected diseases etc. in the fisheries development, which have been pointed by several fish farmers and entrepreneurs of the country.

A general idea for the economics of one hectare of water area, is given below, which has been calculated on the basis of present market rate and will vary from place to place.

Fixed Capital	**Rs.**
1. Reclamation/construction of the pond Rs.30,000 upper limit	
a. Depreciation @ 10%	3000
b. Interest charged @ 14%	4200
2. Cost of nets etc. Rs.5000	
a. Depreciation @ 25%	1250
b. Interest charges @ 14%	700
A. Total fixed cost	9150
Variable Cost	
1. Pond preparation	
a. Lime 200 kg	400-500
b. Manure and fertilizer	
i. Cowdung 50,000kg.	2000
ii. Urea 250 kg.	1000
or	
Ammonium sulphate 800 kg.	
iii. Single super phosphate 500 kg.	1500
2. Cost of 6000 fingerlings @ Rs.60/- per thousand	600
3. Feed 2500 kg. @ Rs.3/- per kg.	7500

4.	a. Fisherman @ Rs.25/- per day	9000
	b. Watchman @ Rs.25/- per day	9000
5.	Water and land taxes	2500
6.	Unforeseen expenditures	2700
B.	Total Variable cost	36,100
7.	Interest charges @14% on Rs.36,100 input cost (C)	5050
D.	Grand Total (A+B+C) 9150+36100+5050	50,300
	Fish production 4000-5000 kg/yr @ Rs.25/kg.	
8.	Total income	1,00,000-1,25,000
9.	Net profit	49,700-74,700
	Cost of production of fish is Rs.10.10/kg.	

Let us consider suitable alternations of land use for betterment of the rural people. While talking of importance of fish farming in rural development, presently the greatest attention is being paid on the integrated approach of fish farming with agriculture and live-stock. The integrated fish farming is based on the principle of recycling of wastes. Live-stock manure (cow, pig, duck, chicken, etc.) and also the nightsoil is used for natural food (plankton) production in the fish culture pond hence the expenditure towards fertilizers and supplementary feed for fish culture is not only curtailed to the barest minimum but also there is economy of space.

Most of the farmers in our country have small fragmented land holdings where modern large scale production technologies with large input requirements do not offer any solution to their problems. These marginal farmers have livestock in the form of few numbers of cattle or a few pigs, a small flock of ducks or chicken and no doubt a surplus family labour where females and children of the house can work all day long.

With these problems and resources various efforts have been made to develop a low cost and self reliant system of farming called integrated system or triple A (Aquaculture-Agriculture-Animal husbandry) system. The substances that are otherwise considered waste are seen as resources and contribute to lower the input as well as increase the production.

Broadly there are three types of integrated fish farming systems widely accepted in our country. These are as following-

1. Fish cum crop/horticulture integration.
2. Fish cum livestock/bird integration.
3. Fish, crop and livestock integration.

The package of practices for rice-cum-fish, fish-cum-duck, fish-cum-poultry, fish-cum-pig and fish-cum-dairy farming have been tested extensively for economic viability and feasibility at the farmers level. Annual fish yields ranging from 3-7 t/ha/yr have been recorded under the different integrated systems, these yields are in addition to the meat and eggs obtained from the livestock raised with fish.

Keeping in mind the demand of integrated farming system a live model was developed at Pantnagar (G.B.Pant Univ. of Agriculture and Technology) on 'Commercialization of Agriculture' where the viable unitary technologies of crop production, horticulture, animal husbandry and fishery were integrated. The model represents a successful technology for marginal farmers forming a self-reliant system for sustainable income.

The whole plot of land i.e. about 3 acres (for marginal farmer) has been divided into 3 equal sectors 1. for aquaculture 2. for horticulture 3. agriculture. The aquaculture sector consists of a earthern pond where adoptable scientific management practices are followed. Papaya and banana trees are planted on the embankments (Fig. 9.1).

Fish production of more than 2t/year is achieved from an acre of water area giving an income of Rs.45,000 per annum. The banana and papaya plants on the pond embankments also add to the income of the farmer. The pond also provides large amount of muleh which serves as good manure for agriculture and horticulture crops. Fertilized water from the ponds can be used for irrigating crops etc.

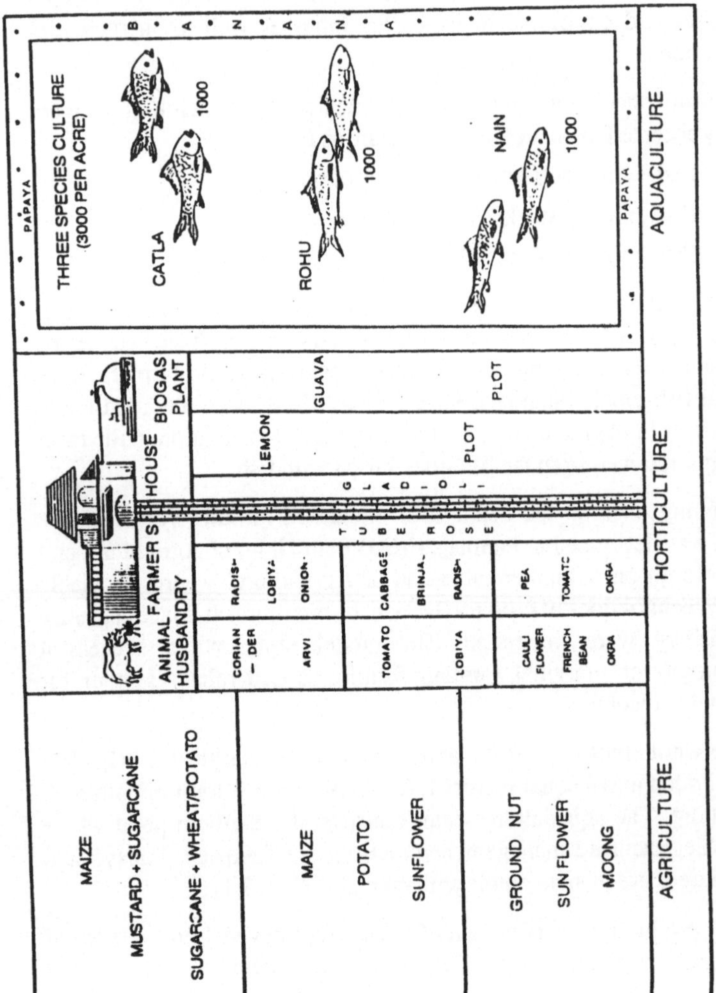

Fig. 9.1: Technology on Commercialisation of Agriculture for Marginal Farmer

Under agriculture sector well tested cash crops are tried in the three sub plots. The crop rotation followed is as under-

Sub plot A : Early sugarcane, $COP_2$11 with intercrop of Toria

Sugar cane returning intercropped with wheat

Sub plot A_2: Maize Potato-Sunflower

Sub plot A_3: Groundnut - Sunflower-Moong

In the horticulture sector the whole plot (1 acre) has been divided into 3 sub plots. In two sub plots.

Guava and citrus fruits are grown-

Sub plot H_1 : Guava

Sub plot H_2 : Citrus fruits

Sub plot H_3 : This plot was further divided into small vegetable growing plots where seasonal vegetables are grown as under.

H_3 : 1 Coriander - Arbi

H_3 : Tomato - Lobiya - Palak

H_3 : Cauliflower-Frenchbean-Lady finger

H_3 4 : Raddish - Onion - Lobiya

H_3 5 : Cabbage - Brinjal - Raddish

H_3 6 : Green pea-Tomato - Lady finger

Farmer's house is also situated in the horticultural sector and in front of the farm house, tube rose and gladuli flower plantation is done. The farmers house is equipped with a small biogas plant and a cowshed for 2 cows. Cow-dung is used for running biogas plants. Biogas provides cooking gas, slurry is used for manuring agricultural and horticultural crops and also pond.

Vegetables and flower beds in the horticultural area can give an income of about Rs.20,000 per annum. In the same way it has been estimated that by following intercropping and rotational pattern of each crops, the area allotted for agriculture can give an income of Rs.15,000 per annum.

Thus a marginal farmer having around 3 acres of land can earn Rs.35,000-40,000 annually and that is sufficient for him to run a small family.

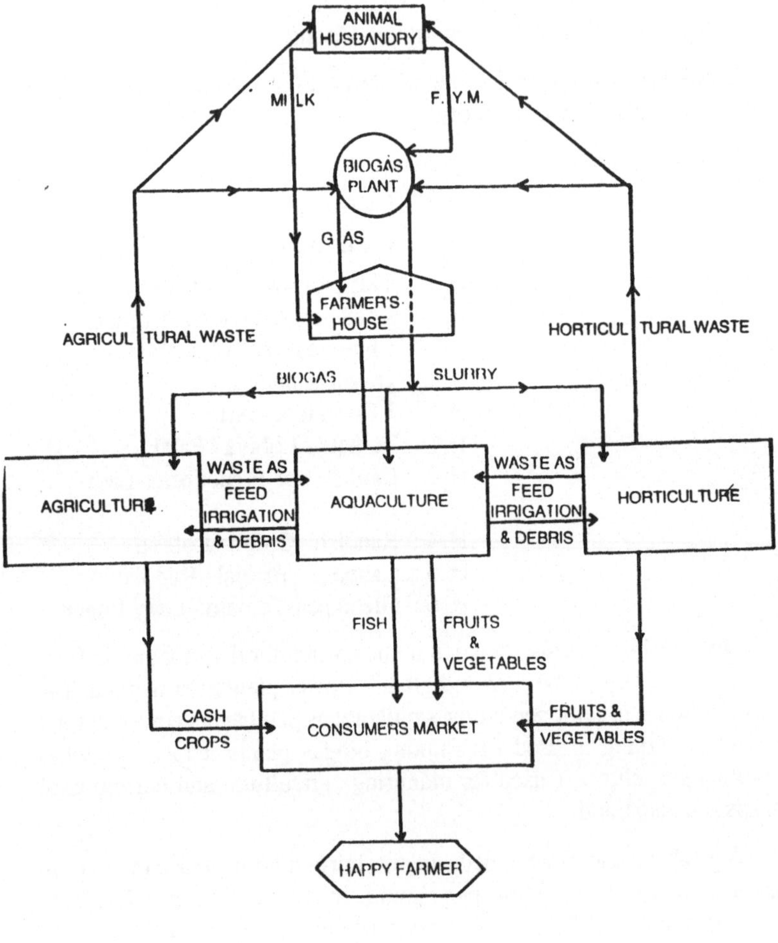

Fig. 9.2: Triple 'A' Recycling System

It is apparent from the foregoing, there is tremendous scope for fish farming in rural areas. Nearly all the tanks, ponds and reservoirs are located in the rural areas. Fish culture should be attempted in the irrigation tanks and canals also. The vast derelict shallow waters in these areas can be reclaimed and utilised for fish culture. If the development of these water bodies is undertaken it is needless to say, this development would greatly ameliorate the socio-economic conditions of people in rural areas.

Demonstration and training to the fish farmers will lead to improvement of fish yield and awareness of the farmers about this lucrative business. Expansion of fish culture in a scientific line by utilization of vast water resources of the country should be order of the day.

Integrated web cycle (Fig. 9.2) in the farm makes it an independent multistructural unit. Such low input technology appears to pay high dividends for Indian farmers by diversified activities. Therefore it is worth to mention that integrated fish farming system has a sound basis for development of integrated rural technology which will enable to fulfil growing needs of the rural people.

Considering the potentiality of fish farming it is hard to think of any other development activity offering such large benefits.

Tips for Bumper Fish Harvest

1. Choose the recommended fish species.
2. Stock suitable size certified seed.
3. Do not delay the pond preparation.
4. Apply recommended quantity of manure and fertilizers through correct methods.
5. Keep the aquatic weeds and insects under control.
6. Maintain the desired water level throughout the year.
7. Do not forget to feed the fish.
8. Adopt proper calendar of management operations.

Index

9 789359 191782

Printed at Repro India Ltd.